EEA Report | No 2/2004

Impacts of Europe's changing climate

An indicator-based assessment

European Environment Agency

Cover design: EEA
Layout: EEA

Legal notice

The contents of this publication do not necessarily reflect the official opinions of the European Commission or other institutions of the European Communities. Neither the European Environment Agency nor any person or company acting on behalf of the Agency is responsible for the use that may be made of the information contained in this report.

Information about the European Union is available on the Internet. It can be accessed through the Europa server (http://europa.eu.int).

Cataloguing data can be found at the end of this publication.

Luxembourg: Office for Official Publications of the European Communities, 2004

ISBN 92-9167-692-6

© EEA, Copenhagen, 2004

Environmental production

This publication is printed according to the highest environmental standards.

Printed by Scanprint a/s
— Environment Certificate: ISO 14001
— Quality Certificate: ISO 9001: 2000
— EMAS registered — licence no. DK- S-000015
— Approved for printing with the Nordic Swan environmental label, licence no. 541 055

Paper
— 100 % recycled and chlorine-free bleached paper
— The Nordic Swan label

Printed in Denmark

541   055
PRINTED MATTER

Do it 100%

European Environment Agency
Kongens Nytorv 6
1050 Copenhagen K
Denmark
Tel. (45) 33 36 71 00
Fax (45) 33 36 71 99
Enquiries: http://www.eea.eu.int/enquiries
Website: http://www.eea.eu.int

Contents

List of maps and graphs

Acknowledgements

This report was prepared by the European Environment Agency's European Topic Centre for Air and Climate Change (ETC/ACC). UBA Berlin (Umweltbundesamt, Federal Environmental Agency) and RIVM (National Institute of Public Health and the Environment, the Netherlands) also contributed financially as partners in ETC/ACC. Thomas Voigt of UBA Berlin and Jelle van Minnen of RIVM coordinated the overall development of the report and were authors of several parts.

Other main authors were Markus Erhard (Atmospheric Environmental Research (MK-IFU), Forschungszentrum Karlsruhe), Marc Zebisch from the Potsdam Institute for Climate Impact Research (PIK), David Viner (Climatic Research Unit — CRU) and Robert Koelemeijer (RIVM). The authors appreciated the advice and comments of Rob Swart of RIVM and of Wolf Garber of UBA Berlin throughout the process.

The EEA project manager was André Jol.

The authors gratefully acknowledge the support of those who contributed text, data, figures and comments:

Joseph Alcamo (University of Kassel, Germany), Michel Bakkenes (RIVM, Bilthoven, the Netherlands), André Berger (EEA Scientific Committee, University of Louvain, Belgium), Gerhard Berz (Munich Re, Munich, Germany), Keith Brander (ICES, Copenhagen, Denmark), Ludwig Braun (Bavarian Academy of Sciences, Munich, Germany), Jerry Brown (IPA, Woods Hole (MA), USA.), Melvin Cannell (CEH, Penicuik, UK), Tim Carter (Finish Environmental Institute, Helsinki, Finland), Philippe Ciais (LSCE, Paris, France), Sophie Condé (ETC on Nature Protection and Biodiversity, Paris, France), Wolfgang Cramer (PIK, Potsdam, Germany), Harry Dooley (ICES, Copenhagen, Denmark), Martin Edwards (SAHFOS, Plymouth, UK), Rune Engeset (NVE, Oslo, Norway), Heidi Escher-Vetter (Bavarian Academy of Sciences, Munich, Germany), Paul Föhn (SLF, Davos, Switzerland), Regula Frauenfelder (WGMS at the University of Zürich, Switzerland), Erik Framstad (NINA, Oslo, Norway), Annette Freibauer (MPI-BGC, Jena, Germany), Karl Gabl (ZAMG, Innsbruck, Austria), Thilo Günther (DWD, Berlin, Germany), Christian Haas (AWI, Bremerhaven, Germany), Wilfried Häberli (WGMS at the University of Zürich, Switzerland), Clair Hanson (CRU, Norwich, UK), Mike Hulme(Tyndall-Centre, UEA, Norwich, UK), Martin Hoelzle (WGMS at the University of Zurich, Switzerland), Hans-Jürgen Jäger (University of Giessen, Germany), Ivan Janssens (University of Antwerpen, Belgium), Gerd Jendritzki (DWD, Freiburg, Germany), Phil Jones (CRU, UEA, Norwich, UK), Frank Kaspar (MPI Hamburg, Germany), Sari Kovats (LSHTM, London, UK), Michael Kuhn (University of Innsbruck, Austria), Bernhard Lehner (University of Kassel, Germany), Günter Liebsch (Technical University of Dresden, Germany), Peter Loewe (BSH, Hamburg, Germany), Grégoire Lois (ETC on Nature Protection and Biodiversity, Paris, France), Christoph Maier (GEUS, Copenhagen, Denmark), Bettina Menne (WHO-ECEH, Rome, Italy), Annette Menzel (Technical University of Munich, Germany), Ranga Myneni (Boston University, Boston (MA), U.S.A), Kristin Novotny (Technical University of Dresden, Germany), Gert-Jan van Oldenborgh (KNMI, de Bilt, the Netherlands), Tim Osborn (CRU, Norwich, UK), Harald Pauli (University of Vienna, Austria), Zbigniew Pruszak (Institute of Hydro-Engineering, Gdansk, Poland), Lars-Otto Reiersen (AMAP, Oslo, Norway), Natalja Schmelzer (BSH, Rostock, Germany),

Klaus Schwarzer (University of Kiel, Germany), Rune Solberg (Euroclim, Oslo, Norway), Johan Ludvig Sollid (University of Oslo, Norway), Oft Stabbetorp (NINA, Oslo, Norway), Wil Tamis (CML, University of Leiden, the Netherlands), Arnold van Vliet (EPN and Wageningen University, Wageningen, the Netherlands), Janet Wijngaard (KMNI, de Bilt, the Netherlands), Sunhild Wilhelms (BSH, Hamburg, Germany), Angelika Wirtz (Munich Re, Munich, Germany).

Finally, EEA acknowledges all who commented on the draft report, in EEA, European Topic Centres, National Focal Points and the European Commission (Directorate General for Environment, Climate change unit).

Summary

Overview

Earth's history has been characterised by many changes in climate conditions. But the extent and the rate of current climate change most likely exceeds all natural variation in the last thousand years and possibly further back in history. There is strong evidence that most of the observed recent warming is attributable to human activities, in particular to emissions of greenhouse gases (GHGs) from burning fossil fuels and land-use changes. Due to ongoing emissions of GHGs, the observed rise in global temperature is expected to continue and increase during the twenty-first century. Climate change already has considerable impacts on the environment, human health and society which are expected to become more severe in future.

As a response to climate change, the United Nations Framework Convention on Climate Change (UNFCCC) has been established. It aims to reduce greenhouse gas emissions and mitigate the effects. Also established are the Kyoto Protocol emission targets for 2008–2012. In addition, EU and national indicative policy targets have been set for future substantial reductions of GHG emissions and for a tolerable projected rise in temperature. To reach such targets, further strategies and policies are needed to achieve more sustainable development in relevant sectors of society (energy, transport, industry, households, agriculture). In addition, strategies will increasingly be required for adapting to the impacts of climate change.

This report presents past trends in Europe's climate, its current state and possible future changes as well as the impacts of climate change on the European environment and society. The report is aimed at the general interested public and decision-makers, especially those who wish to understand which natural systems and societal sectors are most vulnerable to climate change and its impacts.

The main part of the report describes trends in and projections for 22 climate change state and impact indicators. The indicators cover eight categories: the atmosphere; the cryosphere (snow, ice and glaciers); the marine environment; terrestrial ecosystems and biodiversity; water; agriculture; the economy; and human health. The key findings for the 22 indicators are summarised in Table S.1. For almost all indicators, a clear trend exists and impacts are already being observed.

The assessment of climate change and its impacts is still subject to uncertainties and information gaps. The 22 indicators presented in this report illustrate only a small range of the potential consequences of climate change. Other areas are also sensitive to climate change, for instance forestry, water availability, or tourism. Some indicators for these areas have already been developed but have not been included in this report, due to insufficient data availability for Europe or uncertainty in identifying climate change as the cause of changes in these indicators. For some of these areas, information is already available and indicators can be presented in the near future. For others, better knowledge and understanding is needed about the exposure and sensitivity of these systems with respect to climate change.

There is new and stronger evidence that most of the warming observed over the last 50 years is attributable to human activities. Even if society substantially reduces its emissions of greenhouse gases over the coming decades, the climate system would continue to change over the coming centuries.

In order to prevent severe damage to the environment and society, and to ensure sustainable development even under changing climate conditions, adaptation strategies are required. Methods to design and implement adaptation strategies are presented in Chapter 4.

Key findings

1. Atmosphere and climate

Atmospheric indicators show that the concentration of carbon dioxide (CO_2) in the lower atmosphere has increased from its pre-industrial concentration of 280 ppm (parts per million) to its 2003 concentration of 375 ppm. This is the highest level in the last 500 000 years. At the same time, the climate in most parts of the world, including Europe, is warming. The global average temperature has increased by about 0.7 °C and the European average temperature by 0.95 °C in the last hundred years. It is estimated that temperatures will further increase by 1.4–5.8 °C globally and 2.0–6.3 °C in Europe by the year 2100. Precipitation patterns show a more varied picture. Recently, central and northern Europe have received more rain than in the past. In contrast, southern and southeastern Europe have become drier. These changes are projected to continue in the future. In addition, extreme weather events, such as droughts, heatwaves and floods, have increased while cold extremes (frost days) have decreased.

2. Glaciers, snow and ice

One of the most identifiable visual impacts of climate change in Europe can be observed in the cryosphere through the retreat of glaciers, snow cover and Arctic sea ice. Eight out of nine glaciated regions show a significant retreat; the only advancing glaciers are in Norway. From 1850 to 1980, glaciers in the European Alps lost approximately one third of their area and one half of their mass, a trend that is continuing. Even the advances of Norwegian glaciers can be attributed to climate change by increasing winter snowfall. The extent and duration of snow cover across Europe has decreased since 1960. In the Arctic regions of Europe, sea ice has been in decline.

3. Marine systems

The impacts of climate change on the marine environment are covered in this report by assessing the rise in sea level, the sea surface temperature and changes in the marine growing season and species composition. All of these indicators show clear trends. The marine system is mainly affected by an increase in sea surface temperature, especially in isolated basins like the Baltic Sea and the North Sea. This has resulted in an increase in phytoplankton biomass, a northward movement of indigenous zooplankton species by up to 1 000 km within the past few decades, and an increasing presence and number of warm-temperate species in the North Sea. It is estimated that the current rise in sea level of 0.8–3.0 mm/year will continue and intensify by 2.2 to 4.4 times the present values.

4. Terrestrial ecosystems and biodiversity

Terrestrial ecosystems are mainly affected with regard to plant phenology and distribution of plant and animal species. Climate change increased the length of the growing season by 10 days between 1962 and 1995. Northward movement of plant species (induced by a warmer climate) has probably increased species diversity in northwestern Europe, but biodiversity has declined in various other parts of Europe. The survival of different bird species wintering in Europe has increased over the past few decades and is likely to increase further because of the projected rise in winter temperature. The terrestrial carbon uptake of the vegetation has had a positive balance in Europe during the last 20 years. This has led to a removal of some of the atmospheric CO_2 concentration and thus partly mitigated climate change. However, this carbon sequestration will most likely be reduced in future.

5. Water

Annual river discharge is an indicator for both fresh water availability in a river basin and low and high flow events. Annual river discharge has changed over the last decades across Europe. In some regions it has increased, in others, decreased. A part of these changes is attributable to observed changes in precipitation. Annual discharge is expected to decline strongly in southern and southeastern Europe, but increase in northern and northeastern Europe. Therefore, water availability will change over Europe in the coming decades.

6. Agriculture

Climate change affects agriculture in many ways. Increasing atmospheric CO_2 and rising temperatures may allow earlier sowing dates, enhance crop growth and increase potential crop yield. On the other hand, rising temperatures increase the crops' water demand. In combination with changing precipitation patterns, rising temperatures are expected to lead to increasing crop yields in areas with sufficient water supply, to decreasing yields in areas with hot and dry conditions, and to a northward shift of agriculture.

7. Economy

Extreme weather events cause damage to industry, infrastructure and private households. In Europe, a large number of all catastrophic events since 1980 are attributable to weather and climate extremes: floods, storms and droughts/ heatwaves. Economic losses resulting from weather and climate related events have increased significantly during the past 20 years. This is due to wealth increase and more frequent events. Climate change projections show an increasing likelihood of extreme weather events. Thus, a further increase in damage is very likely.

8. Human health

The impact of climate change on human health is evaluated with respect to heatwave-related health problems, tick-borne diseases and flooding. An increase in these impacts has been observed in recent decades and they are projected to escalate further due to projected rises in temperature.

Adaptation

Even if society substantially reduces its emissions of greenhouse gases over the coming decades, the climate system is projected to continue to change over the coming centuries. Therefore, society has to prepare for and adapt to the consequences of some inevitable climate change, in addition to mitigation measures. To prevent or limit severe damage to the environment, society and economies, adaptation strategies for affected systems are required at European, national, regional and local level. The report provides a general framework for adaptation strategies and a number of examples.

Table S.1　Summary of trends and projections of indicators included in this report

Indicators	Key messages
Atmosphere and climate	
Greenhouse gas concentrations	• Due to human activities, the concentration of carbon dioxide (CO_2), the main greenhouse gas, has increased by 34 % compared with pre-industrial levels, with an accelerated rise since 1950. Other greenhouse gas concentrations have also risen as a result of human activities.
	• The total rise in all greenhouse gases since the pre-industrial era amounts to 170 ppm CO_2-equivalent, with contributions of 61 % from CO_2, 19 % from methane, 13 % from CFCs and HCFCs, and 6 % from nitrous oxide.
	• If no climate-driven policy measures are implemented, a further increase to 650–1 215 ppm CO_2-equivalent is projected to occur by 2100.
	• To achieve the EU long-term objective of limiting global temperature rise to 2 °C, global emissions of greenhouse gases need to be reduced substantially from 1990 levels.
Global and European air temperature	• The global average temperature has increased by 0.7 ± 0.2 °C over the past 100 years. The 1990s were the warmest decade in the observational record; 1998 was the warmest year, followed by 2002 and 2003.
	• Europe has warmed more than the global average, with a 0.95 °C increase since 1900. Temperatures in winter have increased more than in summer. The warming has been greatest in northwest Russia and the Iberian Peninsula.
	• The rate of global warming has increased to 0.17 ± 0.05 °C per decade, a value probably exceeding any 100-year rate of warming during the past 1 000 years. The indicative target of no more than 0.1–0.2 °C per decade has already been exceeded or will be exceeded within the next few decades.
	• From 1990 to 2100, the global average temperature is projected to increase by 1.4–5.8 °C and 2.0–6.3 °C for Europe (without policy measures). The 'sustainable' EU target of limiting global temperature increase to no more than 2.0 °C above pre-industrial levels is likely to be exceeded around 2050.
European precipitation	• Annual precipitation trends in Europe for the period 1900–2000 show a contrasting picture between northern Europe (10–40 % wetter) and southern Europe (up to 20 % drier). Changes have been greatest in winter in most parts of Europe.
	• Projections for Europe show a 1–2 % increase per decade in annual precipitation in northern Europe and an up to 1 % per decade decrease in southern Europe (in summer, decreases of 5 % per decade may occur). The reduction in southern Europe is expected to have severe effects, e.g. more frequent droughts, with considerable impacts on agriculture and water resources.
Temperature and precipitation extremes	• In the past 100 years the number of cold and frost days has decreased in most parts of Europe, whereas the number of days with temperatures above 25 °C (summer days) and of heatwaves has increased.
	• The frequency of very wet days significantly decreased in recent decades in many places in southern Europe, but increased in mid and northern Europe.
	• Cold winters are projected to disappear almost entirely by 2080 and hot summers are projected to become much more frequent.
	• It is likely that, by 2080, droughts as well as intense precipitation events will become more frequent.
Glaciers, snow and ice	
Glaciers	• Glaciers in eight out of the nine glacier European regions are in retreat, which is consistent with the global trend.
	• From 1850 to 1980, glaciers in the European Alps lost approximately one third of their area and one half of their mass. Since 1980, another 20–30 % of the remaining ice has been lost. The hot dry summer of 2003 led to a loss of 10 % of the remaining glacier mass in the Alps.
	• Current glacier retreat in the Alps is reaching levels exceeding those of the past 5 000 years.
	• It is very likely that the glacier retreat will continue. By 2050, about 75 % of the glaciers in the Swiss Alps are likely to have disappeared.

Table S.1   Summary of trends and projections of indicators included in this report (cont.)

Snow cover	• The northern hemisphere's annual snow cover extent has decreased by about 10 % since 1966.
	• The snow cover period in the northern hemisphere land areas between 45 °N and 75 °N shortened by an average rate of 8.8 days per decade between 1971 and 1994.
	• Northern hemisphere snow cover extent is projected to decrease further during the twenty-first century.
Arctic sea ice	• The total area of Arctic sea ice has shrunk by more than 7 % from 1978 to 2003.
	• Ice thickness decreased by about 40 % on average from the period 1958–1976 to the period 1993–1997, with large regional variability.
	• The duration of the summer melt season over a large proportion of the perennial Arctic sea ice increased by 5.3 days (8 %) per decade from 1979 to 1996.
	• Projections show a predominantly ice free Arctic Ocean in summer by 2100.
Marine systems	
Rise in sea level	• Sea levels around Europe increased by between 0.8 mm/year (Brest and Newlyn) and 3.0 mm/year (Narvik) in the past century.
	• The projected rate of sea level rise between 1990 and 2100 is 2.2 to 4.4 times higher than the rate in the twentieth century, and sea level is projected to continue to rise for centuries.
Sea surface temperature	• Since the late nineteenth century, the global average sea surface temperature has increased by 0.6 ± 0.1 °C, consistent with the increase in global air temperature.
	• Global ocean heat content has increased significantly since the late 1950s. More than half of the increase in heat content has occurred in the upper 300 metres of the ocean.
	• No European sea shows a significant cooling; the Baltic and North Seas and the western Mediterranean show a slight warming of about 0.5 °C over the past 15 years.
	• It is very likely that the oceans will warm less than the land; by 2100, global sea surface temperature is projected to increase by 1.1–4.6 °C from 1990 levels.
Marine growing season	• Increasing phytoplankton biomass and an extension of the seasonal growth period have been observed in the North Sea and the North Atlantic over the past decades.
	• In the 1990s, the seasonal development of decapods larvae (zooplankton) occurred much earlier (by 4–5 weeks), compared with the long-term mean.
Marine species composition	• Over the past 30 years there has been a northward shift of zooplankton species by up to 1 000 km and a major reorganisation of plankton ecosystems.
	• The presence and number of warm-temperate species have been increasing in the North Sea over the past decades.
Terrestrial ecosystems and biodiversity	
Plant species composition	• Climate change over the past three decades has resulted in decreases in populations of plant species in southern and northern Europe.
	• Plant species diversity has increased in northwestern Europe due to a northward movement of southern thermophilic species, whereas the effect on cold tolerant species is still limited.
	• Projections predict a further northward movement of many plant species. By 2050 species distribution is projected to become substantially affected in many parts of Europe.
	• Globally a large number of species might become extinct under future climate change. Due to non-climate related factors, such as the fragmentation of habitats, extinction rates are likely to increase. These factors will limit the migration and adaptation capabilities needed by species to respond to climate change.

Table S.1   Summary of trends and projections of indicators included in this report (cont.)

Plant species distribution in mountain regions	• Endemic mountain plant species are threatened by the upward migration of more competitive sub-alpine shrubs and tree species, to some extent because of climate change.
	• In the Alps, upward migration has led to an increase in plant species richness in 21 out of 30 summits, whereas it has decreased or remained stable in the other summits.
	• Projected changes in European annual average temperature are outside the tolerance range of many mountain species. These species are projected to be replaced by more competitive shrub and tree species, leading to considerable loss of endemic species in mountain regions.
Terrestrial carbon uptake	• In the period 1990–1998 the European terrestrial biosphere was a net sink for carbon and therefore partly offset increasing anthropogenic CO_2 emissions.
	• Carbon uptake in Europe can be increased by (re-)planting forests and other land management measures. The additional potential storage capacity for the EU in forestry and agriculture is estimated to be relatively small, compared with the agreed targets in the Kyoto Protocol.
	• The projected increase in average temperature is likely to reduce the potential amount of carbon that can be sequestrated in the European terrestrial biosphere in the future.
Plant phenology and growing season	• The average annual growing season in Europe lengthened by about 10 days between 1962 and 1995, and is projected to increse further in the future.
	• Greenness (a measure of plant productivity) of vegetation increased by 12 %, an indicator of enhanced plant growth.
	• The positive effects of temperature increase on vegetation growth (i.e. a longer growing season) are projected to be counteracted by an increased risk of water shortage in mid and especially southern Europe which would adversely affect vegetation.
Bird survival	• The survival rate of different bird species wintering in Europe has increased over the past few decades.
	• The survival rate of most bird species is likely to improve further because of the projected rise in winter temperature.
	• Nevertheless, it is not yet possible to determine what impact this increasing survival will have on bird populations.
Water	
Annual river discharge	• Annual river discharge has changed over the past few decades across Europe. In some regions, including eastern Europe, it has increased, while it has decreased in others, including southern Europe. Some of these changes can be attributed to observed changes in precipitation.
	• The combined effect of projected changes in precipitation and temperature will in most cases amplify the changes in annual river discharge.
	• Annual discharge is projected to decline strongly in southern and southeastern Europe, but to increase in almost all parts of northern and northeastern Europe, with consequences for water availability.
Agriculture	
Crop yield	• The yields per hectare of all cash crops have continuously increased in Europe in the past 40 years due to technological progress, while climate change has had a minor influence.
	• Agriculture in most parts of Europe, particularly in mid and northern Europe, is expected to potentially benefit from increasing CO_2 concentrations and rising temperatures.
	• The cultivated area could be expanded northwards.
	• In some parts of southern Europe, agriculture will be threatened by climate change due to increased water stress.
	• During the heatwave in 2003, many southern European countries suffered drops in yield of up to 30 %, while some northern European countries profited from higher temperatures and lower rainfall.
	• Bad harvests could become more common due to an increase in the frequency of extreme weather events (droughts, floods, storms, hail) and pests and diseases.

Table S.1   Summary of trends and projections of indicators included in this report (cont.)

Economy	
Economic losses	• In Europe, 64 % of all catastrophic events since 1980 are directly attributable to weather and climate extremes: floods, storms and droughts/ heatwaves. 79 % of economic losses caused by catastrophic events result from these weather and climate related events.
	• Economic losses resulting from weather and climate related events have increased significantly during the past 20 years, from an annual average of less than USD 5 billion to about USD 11 billion. This is due to wealth increase and more frequent events. Four out of the five years with the largest economic losses in this period have occurred since 1997.
	• The average number of annual disastrous weather and climate related events in Europe doubled over the 1990s compared with the previous decade, while non-climatic events such as earthquakes remained stable.
	• Climate change projections show an increasing likelihood of extreme weather events. Thus, an escalation in damage caused is likely.

Human health	
Heatwaves	• More than 20 000 excess deaths attributable to heat, particularly among the aged population, occurred in western and southern Europe during the summer of 2003.
	• Heatwaves are projected to become more frequent and more intense during the twenty-first century and hence the number of excess deaths due to heat is projected to increase in the future. On the other hand, fewer cold spells will likely reduce the number of excess deaths in winter.
Flooding	• Between 1975 and 2001, 238 flood events were recorded in Europe. Over this period the annual number of flood events clearly increased.
	• The number of people affected by floods rose significantly, with adverse physical and psychological human health consequences.
	• Fatal casualties caused per flood event decreased significantly, likely due to improved warning and rescue measures.
	• Climate change is likely to increase the frequency of extreme flood events in Europe, in particular the frequency of flash floods, which have the highest risk of fatality.
Tick-borne diseases	• Tick-borne encephalitis cases increased in the Baltic region and central Europe between 1980 and 1995, and have remained high. Ticks can transmit a variety of diseases, such as tick-borne encephalitis (TBE) and Lyme disease (in Europe called Lyme borreliosis).
	• It is not clear how many of the 85 000 cases of Lyme borreliosis reported annually in Europe are due to the temperature increase over the past decades.

1   Introduction

1.1 Purpose and scope of this report

During recent decades, there have been notable changes in the global and European climate. Temperatures are rising, precipitation in many parts of Europe is changing and weather extremes show an increasing frequency in some regions (IPCC, 2001a). According to the UN Intergovernmental Panel on Climate Change (IPCC), 'there is new and stronger evidence that most of the warming observed over the last 50 years is attributable to human activities, in particular to the emission of greenhouse gases' (IPCC, 2001a).

Human induced climate change is expected to continue in the coming decades (IPCC, 2001a), with considerable effects on human society and the environment. The magnitude of the impacts strongly depends on the nature and rate of future temperature increase. Consequences of climate change include an increased risk of floods and droughts, losses of biodiversity, threats to human health, and damage to economic sectors such as forestry, agriculture, tourism and the insurance industry (IPCC, 2001b). In some sectors, new opportunities might occur, depending on the location in Europe. Some of the impacts are already beginning to appear.

This report presents the results of an indicator-based assessment of recent and projected climate changes and their impacts in Europe. The European Topic Centre on Air and Climate Change (ETC/ACC) prepared the report for the European Environment Agency (EEA). The objectives of the report are to:

- present the extent to which climate change and its impacts are already occurring and projected to occur in future;

- enable the assessment of the vulnerability of natural and societal sectors to climate change, and to enable the development of adaptation strategies;

- show the distance to achieving (long-term) climate change targets, in particular the EU indicative target for global temperature;

- raise awareness of how mitigation policies to reduce greenhouse gas emissions can delay or avoid potential adverse impacts of climate change in Europe.

This report focuses mainly on European-wide trends, but adds information on global trends where relevant. Detailed information on regional impacts of climate change is provided in national climate change indicator reports, such as for the UK (Cannell, 2003; Hulme et al., 2002) and for Ireland (Sweeney et al., 2002).

This report is relevant for the general interested public, and for policy- and decision-makers, especially those who wish to understand which impacts of climate change are already noticeable, how these impacts will continue in future, and which natural systems and societal sectors are most vulnerable to climate change. The report builds on a number of more detailed indicator fact sheets, some of which are or will be published separately on the EEA web site.

1.2 Outline

Chapter 2 of this report sets out the background, which helps to understand the need for an assessment of climate change and its impacts. Past and future climate change and the causes of climate change are described. It is shown that climate changes in the past thousands to millions of years were driven by natural forces, whereas the accelerated climate change in the last hundred years is to a large extent attributable to human activities.

The section on climate change policy and sustainable development discusses the policy relevance of climate change and its impacts. The current policy framework of the UN Framework Convention on Climate Change (UNFCCC) and the Kyoto Protocol is explained and indicative policy targets are presented. Further policy strategies are summarised which are aimed at reducing emissions of greenhouse gases or enhancing 'carbon sinks', as well as at adapting to the consequences of climate change. Finally, links to other related environmental policy issues and frameworks (biodiversity, water and human health) are shown.

The main part of the report is Chapter 3. The state of climate change and its impacts in Europe are described by means of 22 indicators, divided into eight different categories:

- Atmosphere and climate

- Glaciers, snow and ice

- Marine systems

- Terrestrial ecosystems and biodiversity

- Water

- Agriculture

- Economy

- Human health.

The indicators present selected and measurable examples of climate change and its impacts, which already show clear trends in response to climate change. The responses of the indicators can be understood as being representative of the more complex responses of the whole category. Furthermore, the results can give an indication of where, to what extent and in which sectors Europe is vulnerable to climate change, now and in the future.

Each indicator is presented in a separate sub-chapter containing a summary of the key messages, an explanation of the relevance of this indicator for the environment, society and policy, and a description of past, recent and future trends.

Chapter 4 stresses the need for adaptation strategies and reviews how these may be set up and how they could help to prevent severe damage from the consequences of climate change.

Finally, Chapter 5 evaluates the difficulties and challenges of attempting assessments of climate change. It explains causes of uncertainties and discusses data availability and quality. It also proposes potential indicators which could broaden future climate impact assessments.

2　Background

2.1　Past and future climate change

Human life cannot exist without the Earth's climate creating suitable environmental conditions for sufficient food, fresh water supply and other essential ecosystem services. Further scientific evidence is becoming available showing that the climate on Earth has changed in recent decades more rapidly than the changes to which human civilisation has adapted in the past. The following section summarises current scientific knowledge about the causes of past and recent climate change.

2.1.1　Natural changes in the climate

Earth's history has shown many changes in climate conditions. Some of these are singular events, resulting in large changes in climate conditions within years or decades. Others show a regular behaviour following different cycles. Most of these other changes occurred over periods of hundreds, thousands or millions of years. They were driven by natural phenomena such as variations in the Earth's orbit around the sun, variations in the Earth axis, fluctuations in the sun's activity and volcanic eruptions. In the past 400 000 years, the climate has shown a periodic cycle of ice ages and warm periods (Figure 2.1). Compared with these variations, the climate of the last 8 000 years has been relatively stable with very small temperature fluctuations (less than 1 °C per century). This stability offered favourable conditions for the development of human society in this period (Petit et al., 1999).

2.1.2　Human induced climate change

Since the beginning of the twentieth century, the Earth's climate has warmed rapidly by about 0.7 °C, with an increase of 0.95 °C in Europe (Climatic Research Unit — CRU, 2003). These changes are unusual in terms of both magnitude and rate of temperature change. The warming exceeds by far all natural climate variations of the last 1 000 years (IPCC, 2001a) (Figure 2.2). The 1990s in particular were the warmest decade in this period (IPCC, 2001a) and the temperature is expected to increase further in the future (see Section 3.2).

Natural causes can explain only a small part of this global warming. There is new and stronger evidence that most of the warming is attributable to human activities, in particular to the emission of greenhouse gases (IPCC, 2001a).

Greenhouse gases have the ability to intercept and re-emit heat which is emitted from the Earth's surface, and thus lead to increases in global temperature. Greenhouse gases are very important for the global climate system. Without natural (pre-industrial) greenhouse gases, global average temperature would be 34 °C lower than it is now, too cold to support human life.

On the other hand, a significant increase in greenhouse gases will lead to a rise in temperature. This may affect natural and societal systems to a degree that could be hard for human society to adapt to.

The main greenhouse gas attributable to human activities is carbon dioxide (CO_2) derived from burning fuels (coal, oil, gas). Other important anthropogenic greenhouse gases include methane (CH_4) from agriculture, nitrous oxide (N_2O) from agriculture and industry, industrial halogenated gases (CFCs and HCFCs) and ozone, which is formed from

Figure 2.1   Reconstructed record of the global average temperature and atmospheric CO$_2$ concentration over the last 400 000 years

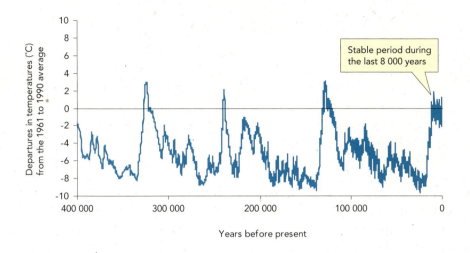

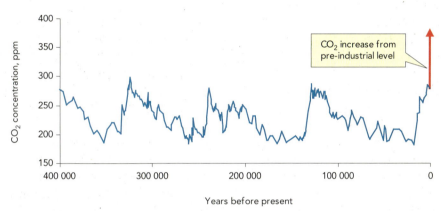

Source: Petit *et al.*, 1999.

compounds emitted by human activities (industry, road transport, households, energy industries).

Anthropogenic emissions have increased the atmospheric concentration of CO$_2$ from 280 ppm (pre-industrial levels, before about 1750) to 375 ppm at present, which exceeds the highest concentration in the last 400 000 years by 70 ppm (Figure 2.1).

2.1.3 Future climate change

The extent of future climate change cannot be known with certainty, since the scientific knowledge of various climate processes is incomplete and socio-economic development, which determines the magnitude of greenhouse gas emissions, is uncertain.

However, there is increasing scientific confidence in the ability of climate models to project the future climate, using projections of greenhouse gas emissions as input. According to these models, the global average surface warming by 2100 will be between 1.4 and 5.8 °C above 1990 levels (Figure 2.2), using a broad range of scenarios of possible socio-economic developments and related greenhouse gas emissions (IPCC, 2001a, see also Section 3.2).

Besides these more or less linear projected trends of future climate, there are additional risks of non-linear or so-called singular events which could be induced by further global warming. The probability that such an event will happen within the next hundred years is relatively low but, if it does occur, the impacts will be extremely high and

Figure 2.2   Reconstructed and measured temperature over the last 1 000 years (northern hemisphere) and projected temperature rise in the next 100 years

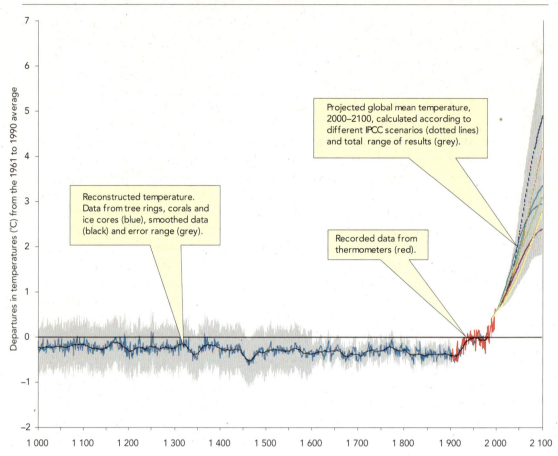

Source: Mann *et al.*, 1999 (last 1 000 years); IPCC, 2001a (projection for the next 100 years).

adaptation would be very difficult. Examples of such potential future singular events are:

- a shutdown of the thermohaline circulation in the North Atlantic (the so called 'North Atlantic Current', also incorrectly referred to as 'Gulf Stream'). This may lead to considerable cooling in northern and western Europe.

- emissions of large amounts of methane from natural gas hydrates in the ocean, deep lakes and polar sediments which could accelerate global warming.

- the disintegration of the West Antarctic Ice Sheet or the melting of the Greenland ice, which could lead to a rise in sea level by several metres.

Due to the very low probability of such events and uncertain scientific knowledge, singular events have not

been considered in this report. This issue may be considered in future reports after more information becomes available (see IPCC, 2001a and WGBU (German Advisory Council on Global Change), 2003a).

2.2   Climate change policy and sustainable development

The continuing and accelerating rate of global climate change and its potentially severe impacts on nature and human society call for policy responses. These responses should mitigate climate change and its impacts as far as possible and help adaptation to the partly inevitable consequences. This section presents climate change policy frameworks and shows links to other policy issues. Some of the policy targets quoted in this chapter are compared to the corresponding climate state and impact indicators in Chapter 3.

2.2.1 Current policy framework

The United Nations opened the Framework Convention on Climate Change (UNFCCC) for signature in 1992 and the convention came into force in 1994. The ultimate objective of the UNFCCC is 'to achieve stabilisation of greenhouse gas concentrations in the atmosphere at a level that would prevent dangerous anthropogenic interference with the climate system. Such a level should be achieved within a time-frame sufficient to allow ecosystems to adapt naturally to climate change, to ensure that food production is not threatened and to enable economic development to proceed in a sustainable manner' (UNFCCC, 1993). By the end of the twentieth century, over 175 states had ratified the convention, indicating that many countries, both industrialised and developing, were convinced that climate change is a serious threat. The EU identified climate change as one of the key environmental concerns in the context of sustainable development (European Parliament and Council, 2002).

To limit climate change and its impacts, it was agreed in 1997 to supplement the Framework Convention with the so-called Kyoto Protocol, which sets quantitative limits for emissions of six greenhouse gases (CO_2, CH_4, N_2O and three groups of fluorinated gases) by developed countries. The target for industrial countries as a whole is a 5 % emission reduction by the 2008–2012 commitment period, compared with the base year (1990 for most countries and for the most important Kyoto gases, except the fluorinated gases). The EU, which at that time had 15 Member States, committed itself under the protocol to reduce its emissions by 8 %. Within this overall target, differentiated emission limitation or reduction targets have been agreed for each Member State under an EU accord known as the

'burden-sharing' agreement. However, the 10 Member States which joined the EU in 2004 keep their individually agreed reduction targets under the Kyoto Protocol, ranging from 6 to 8 % from the base year levels. Countries are allowed to use so-called Kyoto or flexible mechanisms to fulfil their commitments, including project-based joint implementation between developed countries, and clean development mechanisms between developed and developing countries. To some extent also, carbon sinks (ecosystems which can sequestrate carbon, see Section 3.5.3) can count towards the fulfilment of reduction commitments. So far, 120 countries have ratified the Protocol and many of these have adopted national programmes for reducing greenhouse gas emissions. However, Kyoto has not entered into force yet since the emissions of industrialised countries which have ratified the protocol do not reach the threshold of representing 55 % of the base year emissions. If Russia were to ratify, that threshold would be reached. Some industrialised countries, including the US and Australia, have made clear that they will not ratify the Kyoto Protocol.

2.2.2 Long-term policies and sustainable development

The Kyoto Protocol is only a first step towards avoiding 'dangerous anthropogenic interference with the climate system'. Although the Protocol has not yet entered into force, the EU and a number of countries have already expressed their intention to see substantial GHG emission reductions in the longer term. The EU has defined an indicative long-term global temperature target of not more than 2 °C above pre-industrial levels, in addition to a long-term CO_2 stabilisation level of 550 ppm (sixth environment action programme). The German Advisory Council on Global Change has recently proposed the same global temperature target and a CO_2 concentration target of 450 ppm, based on an extensive evaluation of limits to climate change for ecosystems, food production, water availability, economic development and human health (WGBU, 2003a). The United Kingdom (DTI, 2003a, b) adopted a 60 % and Germany aims at a 40 % reduction target of their national emissions (from 1990 levels) by 2050 and 2020, respectively. Likewise, the Netherlands has suggested a 40–60 % emission reduction for western Europe as an indicative target. A recent study (WGBU, 2003b) has proposed reducing global CO_2 emissions from fossil fuels by 45–60 % from 1990 levels by 2050. These targets could be the basis for the global post-Kyoto negotiations, starting in 2005.

To achieve large global reductions in greenhouse gas emissions, both the global share of renewable energy needs to be increased and energy efficiency substantially improved. Technological developments as well as increased research, market penetration strategies and provision of price incentives are necessary. These measures have to be accomplished by a transfer of capital and technology to developing countries. A transformation of global energy systems is essential to provide access to sustainable energy for people in developing countries, which is a UN millennium development goal.

In addition to mitigation strategies, such as emission reduction measures, adaptation to climate change is increasingly receiving attention (see Chapter 4). Such measures are already being developed and implemented in various countries. Within UNFCCC, several climate change funds have been agreed (UNFCCC, 2003).

2.2.3 Climate change and other environmental issues and policies

Since climate change has consequences for nearly all natural and societal systems (see Chapter 3), the issue is considered in the context of other major environmental issues and policy measures such as:

- Biodiversity, which is addressed in the Convention on Biological Diversity (CBD, 2003). This stresses that human activities, including climate change, negatively affect biodiversity. The EU has a goal of halting loss of biodiversity by 2010, which will be influenced by climate change.

- Human health, which can be affected directly (temperature rise) or indirectly (floods) by climate change. Climate change related impacts on human health are addressed by the World Health Organisation (e.g. WHO, 2003).

- Stratospheric ozone depletion (the 'ozone hole'), which can, through atmospheric processes, have an effect on climate change. Strategies to mitigate ozone depletion are addressed by the Montreal Protocol.

- Water, whose future availability can change due to climate change. Water quality (e.g. nitrate levels) and water quantity issues (risk of flooding) are addressed in the EU water framework directive, although climate change is not considered explicitly.

Because of these links, there is an increasing awareness of the need for policies which address these issues simultaneously. Policies to reduce deforestation, for example, could be beneficial both for climate change (deforestation is one of the main sources of CO_2) and biodiversity. Another example is the use of environmental taxes, which tackle different environmental problems simultaneously (EEA, 2003a). Integration of environmental issues in sectoral policies is included in the EU sustainable development strategy (2002). To achieve sustainable development, further integration and harmonisation of policy measures are needed in future.

3 Climate change impacts in Europe

3.1 Introduction

The observed climate change over the last century influences Europe in many ways. It affects natural systems, such as glaciers or ecosystems, as well as societal and economic systems, such as human health and agriculture.

In many cases, climate change is an additional stress factor. Biodiversity, for example, is also affected by factors such as land use changes, overexploitation of natural resources, invasive alien species, and air pollution. But the role of climate change is expected to become more dominant, in particular if the magnitude and rate of climate change is at the higher end of the projected range (IPCC, 2001a, b; WGBU, 2003).

3.1.1 Indicators and vulnerability

Due to the complex interactions among natural and societal systems and the climate system, the impact of climate change cannot be described completely. Instead, changes in well-defined and measurable elements which already show a significant impact of climate change can be used as indicators for changes in the total system. The retreat of glaciers, for instance, can be an indicator for the impact of climate change on snow and ice-related systems. Indicators do not tell the whole story, but they can give clear hints that a system is changing and in which direction or to what extent.

By these means, indicators can help to assess the vulnerability of natural and societal systems to climate change. Vulnerability describes the extent to which a natural or social system is susceptible to sustained damage from climate change, considering the degree of exposure to climate change, the sensitivity of a system and its adaptive capacity (IPCC, 2001b). There is an increasing awareness that Europe is vulnerable to climate change, even though it is probably less vulnerable than developing countries due to its economic capacity to adapt to climate change. Furthermore, indicators can help to show how distant policy targets are. Such targets currently only exist for the global greenhouse gas concentration and global average temperature. But more targets may be defined in future related to the question of what constitutes 'dangerous anthropogenic interference with the climate' (see Section 2.2.1).

For an overview of tools and methods to evaluate the impacts of, and vulnerability and adaptation to, climate change, see UNFCCC (2004).

3.1.2 Selection of indicators

For this report, 22 indicators were selected to describe the state of the climate and the impacts of climate change on various natural and societal systems. These indicators were divided into eight separate categories:

- Atmosphere and climate

- Glaciers, snow and ice

- Marine systems

- Terrestrial ecosystems and biodiversity

- Water

- Agriculture

- Economy

- Human health.

These indicators have been selected because of their measurability, their causal link to climate change, their (policy) relevance, the availability of historical time series, data availability over a large part of Europe (preferably they should cover all of Europe), and their transparency (i.e. they can be easily understood by policy-makers and the general interested audience).

Many other impact indicators (e.g. impact on forestry) were considered for inclusion in the report but were rejected, often because of the difficulty of attributing an observed trend to climate change or due to insufficient data availability (see also Chapter 5) (EEA, 2002b). If more information becomes available in the future, some of these indicators might be reconsidered for inclusion in a future report, to achieve a more comprehensive picture of climate change impacts on the environment and society.

Indicators from existing national indicator sets, such as for the UK (Cannell *et al.*, 1999; Cannell, 2003; Hulme *et al.*, 2002), have been integrated where feasible. Others have been rejected because of missing data for the whole of Europe or because the relevance of these indicators is limited to national issues.

The indicators are part of the full set of indicators that the EEA uses to present the relationships in the 'DPSIR' causality chain (see EEA core set of indicators, EEA, 2003c). This includes the socio-economic driving forces (e.g. energy supply and use), pressures (emissions of greenhouse gases), state of the environment (e.g. the climate), impacts and responses through policies (EEA, 2002a). Indicators on greenhouse gas emissions, removals by carbon sinks and the effectiveness of policies and measures are presented in other reports (European Commission, 2003; EEA, 2003a, b; IEA, 2002) and in National and European communications to UNFCCC (UNFCCC, 2003). They are therefore not addressed here.

3.1.3 Data and information sources for this report

This report uses recorded data and model results to assess past and future climate change and its impact.

While recorded data are a good source for the description of past trends of measurable factors (e.g. temperature), models are needed for the assessment of complex entities which cannot be measured directly (e.g. carbon uptake) and for the assessment of future trends. Models are mathematical formulations of the knowledge about the mechanism of climate change and its impacts. For the model-based assessment of future trends, assumptions about possible changes of the drivers of climate change (e.g. CO_2 emission) are necessary. Such assumptions are called scenarios. Scenarios do not predict a trend but show possible pathways of future development. Scenarios are a common tool to answer questions of the 'what if?' type e.g. 'what would happen to the climate if CO_2 concentrations rose to a level of 650 ppm?'

Data sources used in this report include recent reports on temperature and precipitation (Hadley Centre, 2003) and on climate change and human health (WHO, 2003). In general, the sources of global datasets have been UNFCCC, IPCC, WMO (World Meteorological Organization), WHO (World Health Organization), WGMS (World Glacier Monitoring Service) and the Climatic Research Unit of the University of East Anglia (UK). EU research projects such as ACACIA — A consortium for the application of climate impact assessments (Parry, 2000), CarboEurope (Freibauer, 2002), and the European Phenology Network delivered information on a European scale. Finally, national information about climate change state and impacts indicators has been used where available, e.g. for the UK (Cannell *et al.*, 1999; Cannell, 2003; Hulme *et al.*, 2002) and other countries.

Most of the projections in this report are based on the six emission scenarios published by the UN Intergovernmental Panel on Climate Change (the so-called Special report on emission scenarios — SRES; IPCC, 2000). These scenarios contain plausible estimates of future changes in policies, technologies, land-use, lifestyles, population, economic growth and other issues. These assumptions in turn result in different paths of emissions of greenhouse gases and other pollutants. Emission scenarios are used to assess the consequences of changing emissions for the climate and its impacts. Combining different scenarios and models reduces the uncertainty and thus increases the confidence in a projection.

All information on indicators presented in this report is subject to various types of uncertainties. These can result from gaps in knowledge of climate change processes, insufficient data availability, difficulties in attributing an observed change to climate change and a wide range of possible future socio-economic developments and emissions of greenhouse gases. Uncertainties are briefly addressed in the description of each indicator and explained in more detail in Chapter 5.

3.1.4 Presentation of indicators

The presentation of each indicator comprises three sections:

- key messages that summarise observed and projected trends;

- a relevance section that explains the policy, socio-economic and environmental relevance. It contains information about politically agreed or indicative targets, the possible impacts of climate change, the relevance for other environmental problems and the uncertainties related to the indicator;

- past trends and projections (future trends).

3.2   Atmosphere and climate

3.2.1   Greenhouse gas concentrations

Key messages

- Due to human activities, the concentration of carbon dioxide (CO_2), the main greenhouse gas, has increased by 34 % compared with pre-industrial levels, with an accelerated rise since 1950. Other greenhouse gas concentrations have also risen as a result of human activities.
- The total rise in all greenhouse gases since the pre-industrial era amounts to 170 ppm CO2-equivalent, with contributions of 61 % from CO_2, 19 % from methane, 13 % from CFCs and HCFCs, and 6 % from nitrous oxide.
- If no climate-driven policy measures are implemented, a further increase to 650–1 215 ppm CO_2-equivalent is projected to occur by 2100.
- To achieve the EU long-term objective of limiting global temperature rise to 2 °C, global emissions of greenhouse gases need to be reduced substantially from 1990 levels.

Figure 3.1   Rise of greenhouse gases concentration compared with the year 1750

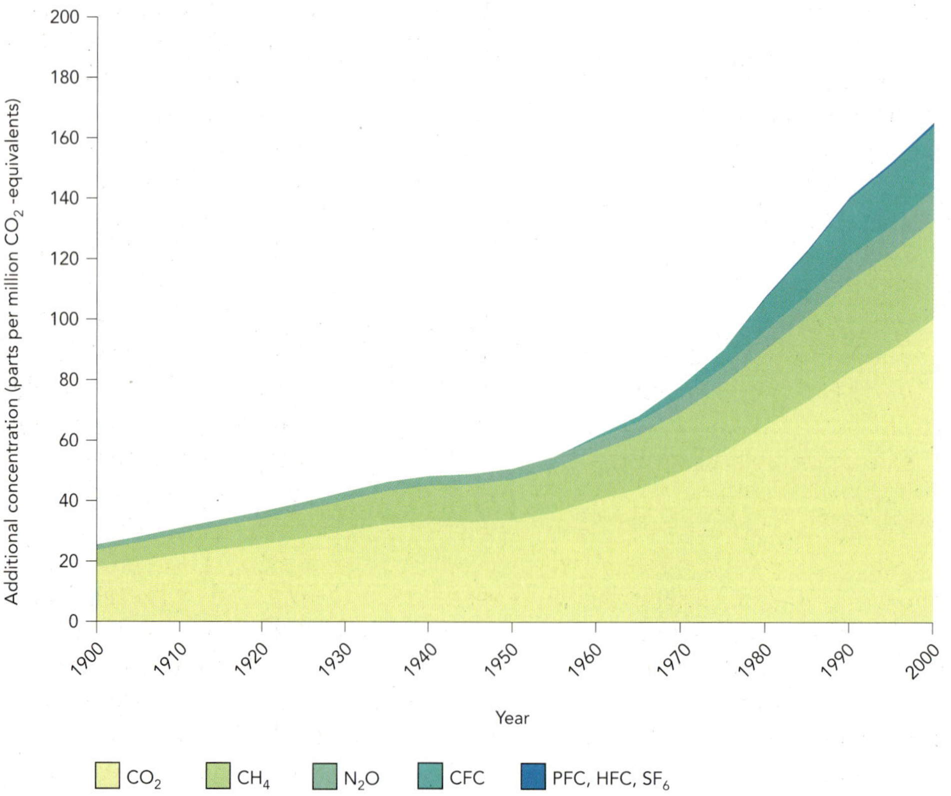

Source: IPCC, 2001a.

Relevance

Emissions of greenhouse gases from human activities are the most important driver of recent climate change. Increases in greenhouse gas concentrations are most likely responsible for most of the observed warming over the last 50 years (IPCC, 2001a).

The ultimate objective of the UN Framework Convention on Climate Change is to stabilise greenhouse gas concentrations in the atmosphere at a level that would prevent dangerous anthropogenic interference with the climate system. The EU has a long-term objective to limit global temperature rise to no more than 2 °C above pre-industrial levels. According to the EU this objective requires a global CO_2 concentration below 550 parts per million (European Parliament and Council, 2002), which is about twice the pre-industrial level of 280 ppm. However, there is considerable scientific uncertainty about whether a limitation to 550 ppm is sufficient to reach the 2 °C target. As a response to new studies, a stricter concentration target of 450 ppm CO_2 has recently been proposed (WGBU, 2003b), in order to stay within the 2 °C temperature rise ceiling.

Source: www.imageafter.com, 2004

The uncertainties in the measurements of GHG concentration are very low (about 1 %). The greater uncertainty in the projection of future trends is mainly due to uncertainties in future emissions and, to a lesser extent, incomplete knowledge of the behaviour of the physical climate system.

Past trends

The concentration of greenhouse gases in the atmosphere increased in the twentieth century due to human activities, mostly related to the use of fossil fuels (e.g. for electric power generation, industry, households and transport), agricultural activities, land-use change (mainly deforestation) and the use of fluorinated gases in industry. The increase has been particularly rapid since 1950. Compared with the pre-industrial era (before 1750), concentrations of carbon dioxide (CO_2) have increased by 34 %, methane (CH_4) by 153 % and nitrous oxide (N_2O) by 17 %. The present concentrations of CO_2 (375 parts per million, ppm) and CH_4 (1 772 parts per billion, ppb) have not been exceeded in the past 420 000 years (for CO_2 likely not even in the past 20 million years); the present N_2O concentration (317 ppb) has not been exceeded in at least the past 1 000 years.

Expressing the concentration of each greenhouse gas as a 'CO$_2$-equivalent' allows a comparison of the different gases. The total concentration of greenhouse gases has increased by 170 ppm CO_2-equivalents since the pre-industrial era (Figure 3.1). Contributions to this rise come from CO_2 (61 %), CH_4 (19 %), N_2O (6 %), and from the halocarbons CFCs and HCFCs (13 %), PFCs, HFCs, and SF_6 (0.7 %). Concentrations of CO_2 and N_2O continue to rise at rates similar to those of the past decades. Concentrations of fluorinated greenhouse gases such as PFCs, HFCs, and SF_6 are rapidly increasing, partly because HFCs are substitutes for ozone depleting gases. In contrast, in the last few years CH_4 concentrations have levelled off, and concentrations of the ozone depleting CFCs and most HCFCs, which are also greenhouse gases, are either decreasing or increasing less rapidly as a result of the ban on their use and production under the Montreal Protocol.

Projections (future trends)

The IPCC has projected different future greenhouse gas concentrations by 2100 (IPCC, 2001a), due to different scenarios of socio-economic,

Figure 3.2   Projected increase of GHG concentration in the atmosphere for four different possible futures

Greenhouse gas concentration (in parts per million CO₂-equivalent)

Year

— A1   — A2   — B1   — B2

Source: IPCC, 2001a.

technological and demographic developments. The scenarios assume no implementation of specific climate-driven policy measures. Under these scenarios, greenhouse gas concentrations are estimated to rise to 650–1 215 ppm CO_2-equivalent by 2100. It is very likely that fossil fuel burning will be the major cause of this increase in the twenty-first century (IPCC, 2001a, b, Figure 3.2).

The IPCC also considered what interventions would be necessary to stabilise atmospheric CO_2 concentration. Total anthropogenic CO_2 emissions

would need to be reduced to below 1990 emissions within a few decades to achieve a stable 450 ppm CO_2 concentration, within about a century to achieve 650 ppm, or within about two centuries to achieve 1 000 ppm CO_2. To reach the EU objective of limiting CO_2 concentrations to no more than 550 ppm requires global emissions of greenhouse gases to be reduced substantially from 1990 levels. A stricter CO_2 stabilisation target of 450 ppm is likely to require a global reduction in emissions of CO_2 of 45 to 60 % by 2050 compared with 1990 levels.

Source: F. Coppin, www.pixelquelle.de, 2004.

3.2.2 *Global and European air temperature*

Key messages

- The global average temperature has increased by 0.7 ± 0.2 °C over the past 100 years. The 1990s were the warmest decade in the observational record; 1998 was the warmest year, followed by 2002 and 2003.
- Europe has warmed more than the global average, with a 0.95 °C increase since 1900. Temperatures in winter have increased more than in summer. The warming has been greatest in northwest Russia and the Iberian Peninsula.
- The rate of global warming has increased to 0.17 ± 0.05 °C per decade, a value probably exceeding any 100-year rate of warming during the past 1 000 years. The indicative target of no more than 0.1–0.2 °C per decade has already been exceeded or will be exceeded within the next few decades.
- From 1990 to 2100, the global average temperature is projected to increase by 1.4–5.8 °C and 2–6.3 °C for Europe. The 'sustainable' EU target of limiting global temperature increase to no more than 2.0 °C above pre-industrial levels is likely to be exceeded around 2050.

Figure 3.3   Observed annual, winter and summer temperature deviations in Europe

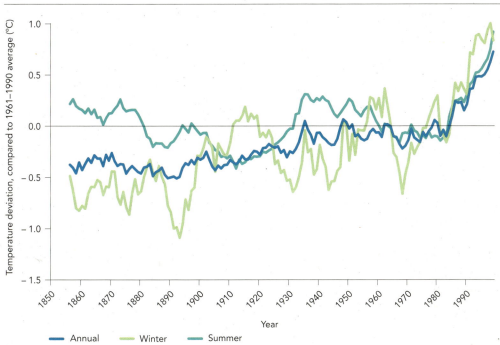

Source: CRU, 2003; Jones and Moberg, 2003.

Relevance

The observed increase in average air temperature, particularly in recent decades, is one of the clearest signals of global climate change. The consequences of rising temperatures include impacts such as increased risk of floods and droughts, biodiversity losses, retreating glaciers and new threats to human health. In addition, temperature increase might damage economic sectors such as forestry, agriculture, tourism and the insurance industry. Some sectors, for example forestry or tourism, may profit from improving environmental conditions, depending on their location. Consequently, sectors might be affected in opposite ways in different parts of Europe. There is mounting evidence that anthropogenic emissions of greenhouse gases are (mostly) responsible for the recently observed increases in average temperature. Natural factors such as volcanoes and sun activity could explain to a large extent the temperature variability up to the mid-twentieth century, but they can explain only a small part of the recent warming (see also Chapter 2). In line with the ultimate objective of the UNFCCC, the EU, in its sixth environmental action programme, has proposed a 'sustainable' target of limiting global average temperature to no more than 2 °C above pre-industrial levels (about 1.3 °C above current global mean temperature). Some studies have proposed an additional 'sustainable' target of limiting the rate of anthropogenic warming, ranging from 0.1 to 0.2 °C per decade (see below).

Temperature has been measured for many decades at many locations in Europe. Of all the indicators in this report, this one has the best coverage across Europe and a low measurement uncertainty. The uncertainty of future temperature is greater due to lack of knowledge of aspects of the climate

system, including climate sensitivity (e.g. resulting temperature rise if CO_2 concentrations are doubled) and seasonal temperature variability (see also Chapter 5).

Past trends

The Earth and Europe have experienced considerable temperature increases in the last 100 years, especially in recent decades. Globally, the increase in the last 100 years was about 0.7 ± 0.2 °C (IPCC, 2001a; CRU, 2003). Within this period, the 1990s were the warmest decade on record; 1998 was the warmest year, followed by 2002 and 2003 (Jones and Moberg, 2003; WMO, 2003).

The rate of global average temperature increase is currently about 0.17 ± 0.05 °C per decade (IPCC, 2001a). Indicative targets restricting the increase to not more than 0.1–0.2 °C per decade have been proposed, based on the limited capabilities of ecosystems to adapt (Rijsberman and Swart, 1990; Leemans

Map 3.1   Annual temperature deviation in Europe in 2003

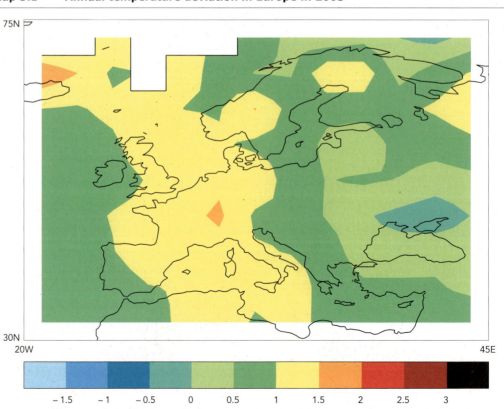

Note: Temperature deviation, relative to average temperature from 1961–1990 (°C).
Source: CRU, 2003; Jones and Moberg, 2003.

and Hootsman, 1998; WBGU, 2003b).
These proposed targets have already
been exceeded or will be exceeded in
the near future.

The temperature increase in Europe
over the last 100 years is about 0.95 °C
(CRU, 2003; Jones and Moberg, 2003),
which is higher than the global average.
The warmest year in Europe was 2000;
the next seven warmest years occurred
in the last 14 years (Figure 3.3). There
is a wide variation in increasing
temperatures across the continent
(Map 3.1). The warming has been
greatest in northwest Russia and the
Iberian Peninsula (Parry, 2000; Klein
Tank *et al.*, 2002). In line with the global
trend, temperatures are increasing in
winter more than in summer (+ 1.1 °C in
winter, + 0.7 °C in summer), resulting in
milder winters and a decreased seasonal
variation (Figure 3.3, Jones and Moberg,
2003).

Source: Glassman, stock.xchng, 2004.

Projections (future trends)

The projected temperature increase
between 1990 and 2100 is likely to be
in the range of 1.4–5.8 °C for the global

Map 3.2   Projected temperature changes in Europe up to 2080

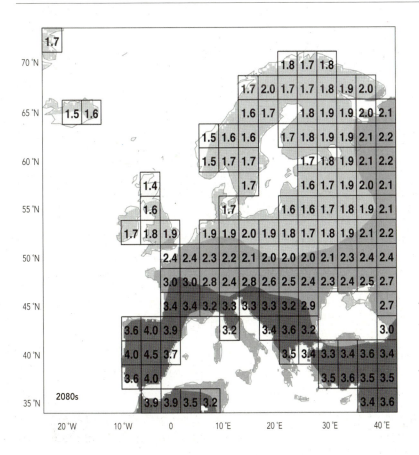

Note: Temperature change (⁰C). Relative to average temperature in the period 1961–1990.
Intermediate ACACIA scenario in a broad range of possible future emissions.
Source: IPCC, 2001b; Parry *et al.*, 2000.

mean (IPCC, 2001a, Map 3.2) and 2–6.3 °C for Europe (Parry, 2000). This range results from potential different pathways of technological, demographic and economic development (leading to different emissions), and is due to uncertainties related to the climate system's response to changing concentrations of greenhouse gases. Looking at the projected range, the EU 'sustainable' target of limiting global average warming to not more than 2 °C above pre-industrial levels might be exceeded between 2040 and 2060.

Within Europe, the warming is estimated to be greatest over southern countries (Spain, Italy, Greece) and the northeast (e.g. western Russia) and less along the Atlantic coastline (Map 3.2). In southern Europe, especially, this may have severe consequences such as increasing drought stress, more frequent forest fires, increasing heat stress and risks for human health. The European trend that winters will warm more rapidly than summers will continue (with the exception of southern Europe).

3.2.3 *European precipitation*

Key messages

- Annual precipitation trends in Europe for the period 1900–2000 show a contrasting picture between northern Europe (10–40 % wetter) and southern Europe (up to 20 % drier). Changes have been greatest in winter in most parts of Europe.

- Projections for Europe show a 1–2 % increase per decade in annual precipitation in northern Europe and an up to 1 % per decade decrease in southern Europe (in summer, decreases of 5 % per decade may occur). The reduction in southern Europe is expected to have severe effects, e.g. more frequent droughts, with considerable impacts on agriculture and water resources.

Map 3.3   Annual precipitation changes in Europe for the period 1900–2000

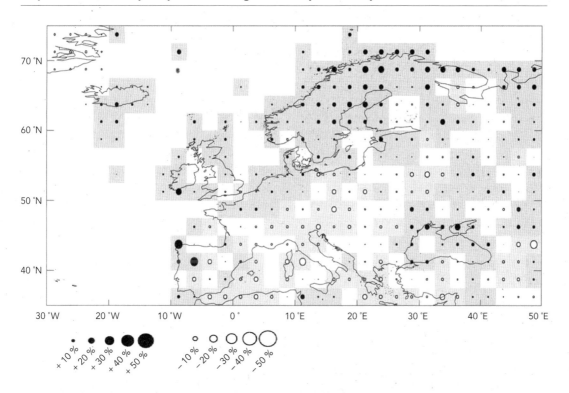

Note: Unit: Percentage change per century. Black circles show areas getting wetter and white circles show areas getting drier. Circle size is related to the magnitude of the trend. Shaded trends are significant at 90 %.
Source: IPCC, 2001b; Parry, 2000.

Relevance

Precipitation includes rain, snow and hail. Changes in average precipitation can have potentially far-reaching impacts on ecosystems and biodiversity, agriculture (food production), water resources and river flows. Changes in precipitation patterns over the year can lead to more flooding in some regions or seasons and more droughts in others, more frequent land slides and soil erosion. Floods and droughts can even occur in the same region in different seasons of the same year (e.g. a region may be exposed to drought in spring and summer and be flooded in autumn).

Data coverage of precipitation across Europe is reasonable and has a low uncertainty. However the data on trends are more uncertain because measurement techniques changed during the twentieth century. Projections of future precipitation have a greater uncertainty. This applies particularly to projections of regional precipitation patterns and seasonal distribution.

Past trends

As an average over all global land areas, annual precipitation increased by 2 % between 1900 and 2000. In Europe, this increase was considerably larger (IPCC, 2001a; Parry, 2000; Klein Tank *et al.*, 2002). The European trend shows a significant difference between seasons and a contrasting picture across the continent (Map 3.3). Annual precipitation increased over northern Europe by 10–40 % in the period 1900–2000, whereas parts of southern Europe experienced a 20 % precipitation decrease (IPCC, 2001a; Klein Tank *et al.*, 2002; NOAA, 2001). The seasonal precipitation pattern shows more pronounced trends across the European continent than the annual changes. In the winter season, especially, southern and eastern Europe became drier (Romero *et al.*, 1998), while many parts of northwestern Europe became wetter (Parry, 2000). The changes in winter precipitation can partly be linked to

specific weather conditions, such as stronger western winds over northern Europe, bringing in more clouds. Summer precipitation shows a decrease of about 10 % in the Mediterranean and central Scandinavian countries.

Projections (future trends)

Uncertainties are high in projected (regional) precipitation change, resulting in a considerable range of projections, although scientific confidence in the ability of climate models to estimate future precipitation has gradually increased. Global average (land and ocean) precipitation is projected to increase by 2–7 % between 1990 and 2100 (IPCC, 2001a). The range is due to uncertainties within the climate models and differences in emission scenarios. Projections for Europe show more annual precipitation for northern Europe (an increase of 1–2 % per decade) and decreasing trends across southern Europe (maximum – 1 % per decade). In winter, most of Europe is likely to become wetter (1–4 % per decade) with some exceptions, notably in the Balkans and Turkey. Despite this increase, the amount of snow is projected to decline due to rising temperatures. In summer, northern Europe might become wetter (up to 2 % per decade), whereas southern Europe may become up to 5 % drier

Source: stock.xchng, 2004.

per decade (Map 3.4) (Parry, 2000; IPCC, 2001b). The latter could have severe impacts on agriculture and water resources as moisture availability is often already limited, especially in summer.

Map 3.4   Projected change in summer precipitation in Europe up to 2080

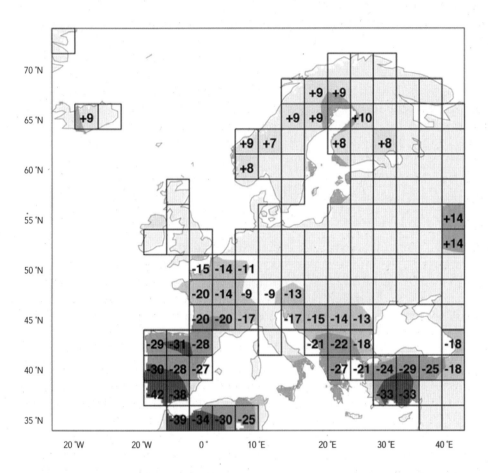

Note: Summer precipitation change (%). Relative to average precipitation in the period 1961–1990. Intermediate ACACIA scenario in a broad range of possible future emissions.
Source: IPCC, 2001b; Parry *et al.*, 2000.

3.2.4 Temperature and precipitation extremes

Key messages

- In the past 100 years the number of cold and frost days has decreased in most parts of Europe, whereas the number of days with temperatures above 25 °C (summer days) and of heatwaves has increased.
- The frequency of very wet days has significantly decreased in recent decades in many places in southern Europe, but increased in mid and northern Europe.
- Cold winters are projected to disappear almost entirely by 2080 and hot summers are projected to become much more frequent.
- It is likely that, by 2080, droughts as well as intense precipitation events will become more frequent.

Map 3.5   Change in frequency of summer days in Europe between 1976 and 1999

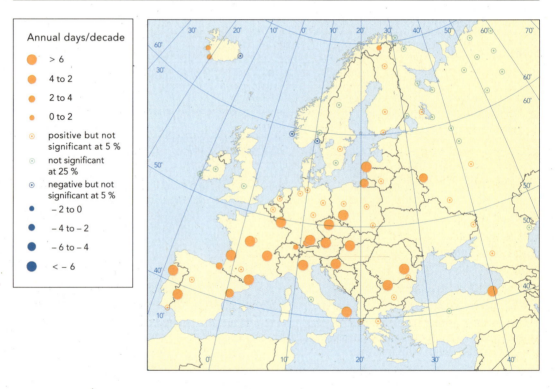

Note: Summer days are defined as days with temperature above 25 °C.
Source: Klein Tank *et al.*, 2002.

Relevance

Temperature and precipitation extremes, such as heatwaves, summer droughts or heavy rain, can have severe impacts on ecosystems and society. Impacts include extreme events like river floods as well as negative effects on ecosystem services such as biodiversity, agriculture, water availability and human health. These negative effects can cause very heavy economic losses, e.g. the droughts of 1999 caused losses of more than EUR 3 billion in Spain (EEA, 2004). The attribution of climate change to these extremes is still uncertain because of a lack of accurate data and scientific understanding of the climate system. Only recently, some first assessments have been published showing that extremes are changing more than the average climate (e.g. Schär *et al.*, 2004). Such information provides an indication of what may increasingly happen in the future.

Map 3.6   Changes in frequency of very wet days in Europe between 1976 and 1999

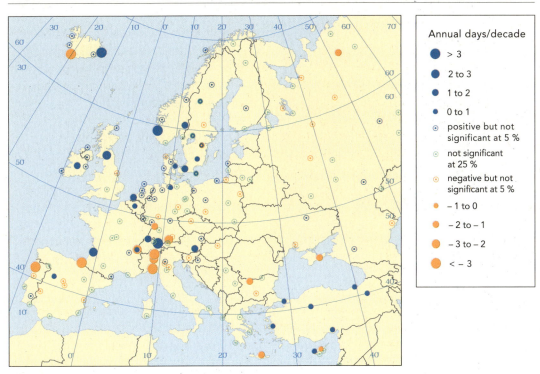

Annual days/decade	
●	> 3
●	2 to 3
●	1 to 2
●	0 to 1
⊙	positive but not significant at 5 %
⊙	not significant at 25 %
⊙	negative but not significant at 5 %
●	– 1 to 0
●	– 2 to – 1
●	– 3 to – 2
●	< – 3

Note: Very wet days are defined as days with precipitation above 20 mm.
Source: Klein Tank *et al.*, 2002.

Past trends

In the past three decades summer days (days with temperatures above 25 °C) and heatwaves have become more frequent (Map 3.5). The most severe changes have been observed in western and southern parts of the continent. At the same time, the number of frost days has decreased even more, due to a greater warming in winter than in summer (Jones *et al.*, 1999; Klein Tank *et al.*, 2002). The tendency towards milder winters in Europe in the last 10–20 years is partly caused by stronger westerly circulation in winters, consistent with a positive phase of the North Atlantic Oscillation (NAO) (Hurrell, 1996). It is scientifically uncertain whether the current trend towards the positive phase of the NAO is part of the human-induced climate change signal or is natural climate variability. It is likely to be a combination of both (Gillett *et al.*, 2003).

Furthermore, the number of wet extremes has increased in Europe in recent decades (Map 3.6, Klein Tank *et al.*, 2002). In many regions (including parts of Russia), the trend in precipitation extremes is more pronounced than the trend in average precipitation. A consistent change over time has occurred for many of the precipitation extremes since 1976. An increase has been observed in the number of very wet days in central and northern Europe, whereas decreases have been observed in parts of southern Europe.

Source: stock.xchng, 2004.

Source: J. W. Joensen, Novolysio, 2000.

In the past decade Europe has been affected by three remarkable weather extremes. The summers of 1995 and 2003 were extremely hot throughout most parts of Europe. In contrast, 2002 was very wet and saw extreme flooding in central Europe. Compared with the historical frequency of extreme weather events, there is some indication that the accumulation of such extreme events in recent decades is uncommon (IPCC, 2001a; Schär *et al.*, 2004).

There is mounting evidence that these changes in the frequency and extent of climate extremes are likely to be caused by a shift of the mean climate to more extreme conditions, and more and stronger deviations from this mean (IPCC, 2001a). Schär *et al.* (2004) showed that the extremely hot summer of 2003 could most likely be explained only by an extension of the climate variability.

Projections (future trends)

Cold winters (which occurred once every 10 years from 1961 to 1990) are likely to become rare and will almost entirely disappear by 2080. In contrast, by 2080 nearly every summer in many parts of Europe is projected to be hotter than the 10 % hottest summers in the current climate. Under high emission scenarios every second summer in Europe will be as hot or even hotter than 2003 by the end of the twenty-first century (Luterbacher *et al.*, 2004). In southern Europe, these changes are projected to occur even earlier (in Spain by the 2020s) (Parry, 2000).

Changes are also projected for precipitation extremes in Europe, but the uncertainties remain high. It is likely that the frequency of both intense precipitation events and summer drought risk will increase (Parry, 2000; Klein Tank *et al.*, 2002). The first could lead to increased flood events, while the latter could have severe consequences for agriculture, water resources and the frequency of forest fires in southern Europe.

3.3 Glaciers, snow and ice

3.3.1 Glaciers

Key messages

- Glaciers in eight out of the nine European glacier regions are in retreat, which is consistent with the global trend.
- From 1850 to 1980, glaciers in the European Alps lost approximately one third of their area and one half of their mass. Since 1980, another 20–30 % of the remaining ice has been lost. The hot dry summer of 2003 led to a loss of 10 % of the remaining glacier mass in the Alps.
- Current glacier retreat in the Alps is reaching levels exceeding those of the past 5 000 years.
- It is very likely that the glacier retreat will continue. By 2050, about 75 % of the glaciers in the Swiss Alps are likely to have disappeared.

Figure 3.4   Cumulative net balance of glaciers from all European glacier regions

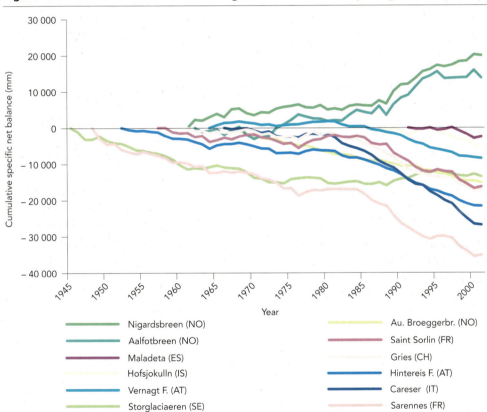

Nigardsbreen (NO)	Au. Broeggerbr. (NO)
Aalfotbreen (NO)	Saint Sorlin (FR)
Maladeta (ES)	Gries (CH)
Hofsjokulln (IS)	Hintereis F. (AT)
Vernagt F. (AT)	Careser (IT)
Storglaciaeren (SE)	Sarennes (FR)

Source: Frauenfelder, WGMS, 2003.

Relevance

About 54 000 km² of Europe (excluding Greenland) is covered by glaciers. 68 % are located on the Svalbard islands, 21 % on Iceland, 6 % in Scandinavia and 5 % in the Alps and the Pyrenees. Mountain glaciers are particularly sensitive to changes in the global climate because their surface temperature is close to the freezing/melting point. Changes in glaciers over time in terms of mass, volume, area and length provide some of the clearest signals of climate change in nature. Glaciers are therefore considered as key indicators for the early detection of global greenhouse related warming trends (IPCC, 2001a).

Vernagt glacier (Austria)
Source: Weber; BAdW/kfG; 1985, 2000.

Past trends

The dynamics of glaciers can be represented by well-explored glaciers in each of the nine European glacier regions (Figure 3.4). In nearly all of the European glacier regions, a general loss of glacier mass has been observed after the so-called 'glacier high-stand' in the middle of the nineteenth century. In the Alps, glaciers lost one third of their surface and one half of their mass between 1850 and 1980. Now there is evidence of a worldwide accelerated trend towards a loss of glacier volume and area since the end of the 1980s (Dyurgerov, 2003). Since 1980, Alpine glaciers, for instance, have lost about 20–30 % of their remaining ice. The hot dry summer of 2003 led to a loss of 10 % of the remaining glacier substance in the Alps (Haeberli, 2003).Current glacier retreat in the Alps is reaching levels exceeding those of the past 5 000 years (IPCC, 2001a).Most of the European glaciers are retreating, losing mass and extent because of warmer summers and a lack of snowfall in the summer season. Only Norwegian coastal glaciers are expanding and gaining mass due to increased snowfall in winter in this region, which is also an effect of climate change (IPCC, 2001a).

Projections (future trends)

Glaciers and permanently frozen mountain areas are among the world's most vulnerable regions to continuing or accelerating climate change. Model studies of individual glaciers indicate a continuous and accelerated retreat induced by global warming in the future (Wallinga and van de Wal, 1998; Haerberli and Beniston, 1998). It is likely that, by 2035, one half — and, by 2050, as much as three quarters — of the present-day glaciers in Switzerland will have disappeared (Maisch and Haeberli, 2003).

The ongoing retreat of glaciers will adversely affect summer skiing in glacier regions and therefore reduce tourism and its economic benefits in these regions (Bürki *et al.*, 2003). Furthermore, it might have adverse impacts on regional water resources.

With a diminished ice mass, the annual melt water and therefore the contribution to river flow and sea level rise is decreasing in the long term. During the melting process, there is an increase in the number of hazardous incidents such as breaking glacier lakes, falling ice and landslides. Glacier retreat affects tourism and winter sports in the mountains and reduces the attractiveness of mountain landscapes. The changes in the water cycle are leading to a reduced supply of drinking water, weakening irrigation facilities and curbing the generation of hydropower. An increased number of hazardous incidents might cause more damage to infrastructure.The uncertainty of observed trends is low, but there is greater uncertainty surrounding projected future glacier retreat. This is due to incomplete knowledge of the climate system, including snowfall during summer in mountain areas.

3.3.2 Snow cover

Key messages

- The northern hemisphere's annual snow cover extent has decreased by about 10 % since 1966.
- The snow cover period in the northern hemisphere land areas between 45 °N and 75 °N shortened by an average rate of 8.8 days per decade between 1971 and 1994.
- Northern hemisphere snow cover extent is projected to decrease further during the twenty-first century.

Figure 3.5   Deviations of monthly snow cover extent over the northern hemisphere lands (including Greenland)

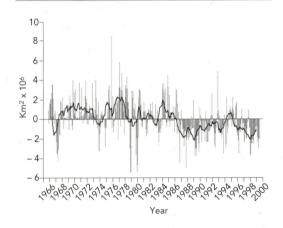

Note: Deviations are compared to a 30 year average. 12-month running means (solid curve).
Source: IPCC, 2001a.

Figure 3.6   Deviations of seasonal snow cover (solid curve) versus deviations of temperature (dashed curve)

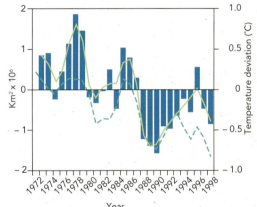

Note: Deviations are compared to a 30 year average.
Source: IPCC, 2001a.

Relevance

Snow cover is an important feedback mechanism in the climate system. The extent of snow cover depends on the climate e.g. on temperature and solar radiation. However it also influences the climate and climate related systems by affecting the reflection of radiation, the thermal insulation, hydrology and length of the growing season. A decrease in snow cover reduces the reflection of solar radiation, contributing to accelerated climate change.

Over 30 % of the Earth's land surface is seasonally covered by snow. The geographic extent of snow cover of the northern hemisphere land (NHL) varies seasonally, reaching a maximum of approximately 46 million km² (about 50 % of the NHL) in January and February, and a minimum of about 4 million km² in August (Armstrong and Brodzik, 2001).

Snow cover affects river discharge, vegetation (thermal insulation) and wildlife (migration patterns). Snow cover retreat has an adverse effect on snow sports and winter tourism as well as on the generation of hydroelectric power based on melt water. On the other hand, snow cover retreat might reduce complications in winter road and rail maintenance, and improve transport.

Source: M. Zebisch, 2004.

Snow data resulting from remote sensing since the mid 1960s and terrestrial measurements are available with low uncertainty. The uncertainty in projected future snow cover is greater due to incomplete knowledge of the climate system and of non-climatic parameters, e.g. landscape properties such as topography and vegetation cover.

Past trends

Satellite records (Figures 3.5 and 3.6) indicate that the northern hemisphere's annual snow cover extent has decreased by about 10 % since 1966 as a consequence of higher land temperatures (IPCC, 2001a). This change mainly affects the snow cover in spring and summer, which has significantly decreased since the mid-1980s over both the Eurasian and American continents (Robinson, 1997, 1999). The duration of the snow-free period in northern hemisphere land areas between 45 °N and 75 °N increased at an average rate of 8.8 (± 1.7) days per decade between 1971 and 1994 (Dye, 1997). Long-term changes of snow cover in Europe have also been documented for Switzerland

(Laternser, 2001). Here, high altitudes show only slight changes while mid and low altitudes show greater changes. The shorter snow cover duration in the Alps is caused by earlier melting of snow in spring rather than by later snowfalls in autumn. The mean snow depth and the duration of snow cover in the Swiss Alps during the observation period 1931–1999 show a gradual increase until the early 1980s, with insignificant interruptions during the late 1950s and early 1970s, followed by a reduction towards the end of the century.

Projections (future trends)

It is estimated that, as global warming proceeds, regions currently receiving snowfall will increasingly receive precipitation in the form of rain. For every 1 °C increase in temperature, the snowline rises by about 150 metres. As a result, less snow will accumulate at low elevations. In contrast, there could be greater snow accumulation in regions above the freezing line due to increased snowfall in some of these regions (IPCC, 2001a). Climate models show that in the future the European Alps and Pyrenees are likely to experience milder winters with more precipitation, and hotter, drier summers (Beniston et al., 1995). These conditions are likely to reduce snow cover on the mountains since, in most temperate mountain regions, the snow temperature is close to the melting point and therefore very sensitive to changes in temperature.

Snowfall in lower mountain areas will become increasingly unpredictable and unreliable over the coming decades (Bürki et al., 2003). As a consequence, nearly half of all ski resorts in Switzerland, and even more in Germany, Austria and the Pyrenees, will face difficulties in attracting tourists and winter sport enthusiasts in the future.

3.3.3 Arctic sea ice

Key messages

- The total area of Arctic sea ice shrank by more than 7 % between 1978 and 2003.
- Ice thickness decreased by about 40 % on average from the period 1958–1976 to the period 1993–1997, with large regional variability.
- The duration of the summer melt season over a large proportion of the perennial Arctic sea ice increased by 5.3 days (8 %) per decade from 1979 to 1996.
- Projections show a predominantly ice free Arctic Ocean in summer by 2100.

Figure 3.7   Monthly deviations of Arctic sea ice extent

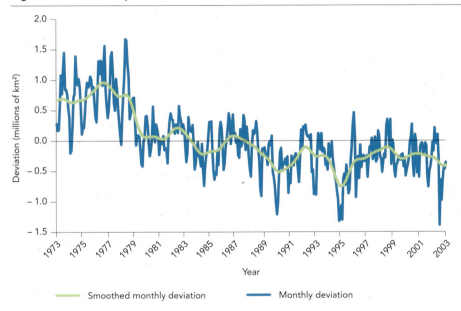

Note: Deviations relative to the monthly average of the 30 years period.
Source: N. Rayner, UKMO, 2004.

Figure 3.8   Regional changes of mean sea ice draft in the Arctic

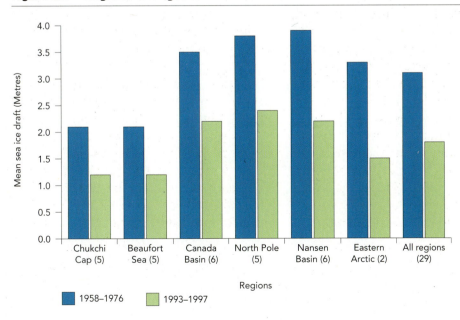

Note: Sea ice draft is the subsurface fraction of the ice thickness.
Source: Rothrock *et al.*, 1999.

Relevance

Arctic sea ice is an important part of the global climate system. Its variability affects the reflection of radiation and the heat exchange between ocean and atmosphere. Arctic sea ice modifies the stratification of the upper layer of the Nordic seas and the thermohaline circulations, e.g. the North Atlantic Current.

Observed changes in the extent of Arctic sea ice provide early evidence of global climate warming.

Long-term trends in the extent and thickness of the ice are a result of changes in atmospheric circulation regimes and oceanic drift patterns, as well as in regional and global temperature and precipitation.

Arctic sea ice is of great relevance to biophysical and socio-economic behaviour in the Arctic and its surroundings. Shrinking sea ice endangers the biological habitats of polar bears, seals and walruses, but increases the CO_2-storing capacity of the ocean by causing more cold open oceanic water. A reduction in sea ice area affects the habitats of indigenous people, impacts on fishery and facilitates marine transportation. Furthermore, it weakens the protection of coasts against severe weather and increases erosion, inundation, the threat to building structures and the dispersal of water pollutants.

Data on sea ice extent result mostly from remote sensing systems and have a low uncertainty. However, because data on ice thickness are often measured by upward-looking sonar on military submarines, access to these data is limited and knowledge gaps exist. Satellite missions, such as 'Cryosat' starting in the near future, will improve data availability. Uncertainty in the attribution of climate change to Arctic ice cover results in uncertainty in projected sea ice changes.

Past trends

Warming in the Arctic has been greater than in any other part of the world, with a 5 °C air-temperature increase in the

Source: H. Bäsemann, 2004.

twentieth century over extensive land areas (IPCC, 2001). During the first half of the twentieth century, the Arctic ice extent remained essentially constant in all seasons. Around 1950, the summer minimum extent began to shrink while the winter maximum remained unchanged. From around 1975 onwards, the winter maximum also began to shrink.

Satellite observations have shown that the total area of Arctic sea ice continuously decreased by at least 7.4 % over the period from 1978 to 2003 (Johannessen et al., 2002, 1995; Bjørgo et al., 1997, Cavalieri et al., 1997; WMO, 2002), with a record low in September 2002 (Johannessen et al., 2002) (Figure 3.7). Satellite data also show a clearly negative trend in the sea ice concentration (surface percentage of the Arctic ocean covered by ice) since the early 1990s (Loewe, 2002).

Although there is considerable regional variability in ice thickness, the average thickness in the different Arctic regions decreased by 40 % from the period 1958–1976 to 1993–1997 (Rothrock et al., 1999) (Figure 3.8). The late summer sea ice thickness in the Arctic Transpolar Drift decreased by about 20 % in the decade 1991–2001 (Haas, 2004). Another recently published paper (Overland et al., 2004) supports the trend of shrinking and thinning Arctic sea ice. A further detailed analysis on impacts of climate change

in the Arctic region will be presented by the Arctic Climate Impact Assessment (ACIA), due at the end of 2004.

Over a large proportion of the Arctic sea ice, the duration of the summer melt season (days with temperatures above zero) increased by 5.3 days (8 %) per decade from 1979 to 1996 (IPCC, 2001a).

The observations of Arctic sea ice retreat in the past decades are well simulated by climate models. Both the observed and modelled trends are much larger than would be expected from natural variations. This indicates that the observed decrease in sea ice extent since 1950 is due to the observed warming in the Arctic region, a consequence of direct and indirect impacts of the enhanced climate change (Moritz et al., 2002).

Projections (future trends)

Global climate model simulations of Arctic sea ice estimate that future global warming will cause a decrease in the maximum ice thickness of about 0.06 metres per °C and an increase in open water duration of about 7.5 days per °C (IPCC, 2001).

Sea ice extent by 2050 might be about 80 % less than in the mid-twentieth century, and might disappear in summer by the end of this century (Johannessen, 2002).

3.4 Marine systems

3.4.1 Rise in sea level

Key messages

- Sea levels around Europe increased by between 0.8 mm/year (Brest and Newlyn) and 3.0 mm/year (Narvik) in the last century.
- The projected rate of sea level rise between 1990 and 2100 is 2.2 to 4.4 times higher than the rate in the twentieth century, and sea level is projected to continue to rise for centuries.

Map 3.7   Change of sea level at selected stations in Europe from 1896 to 1996

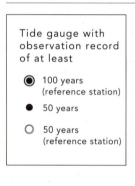

Note: Data are corrected for post-glacial processes.
Source: Liebsch *et al.*, 2002.

Relevance

Sea level rise is an important indicator of climate change, with great relevance in Europe for flooding, coastal erosion and the loss of flat coastal regions. Globally, even entire (island) states are threatened. Rising sea level increases the likelihood of storm surges, enforces landward intrusion of salt water and endangers coastal ecosystems and wetlands.

Apart from natural ecosystems, coastal areas often feature important managed ecosystems, economic sectors, and major urban centres. Thus, a higher flood risk increases the threat of loss of life and property as well as of damage to protection measures and infrastructure, and might lead to an increasing loss of tourism, recreation and transportation functions.

Sea level data have been gathered for upwards of 230 years at different places on European coasts and have a low uncertainty. The range of projected sea level rise results mainly from the range

Figure 3.9   Sea level rise at selected European gauge stations

Stockholm

Esbjerg

Brest

Marseille

Note: Data are corrected for postglacial processes.
Source: Liebsch *et al.*, 2002.

of different emission scenarios, and to some extent from uncertainty regarding the different physical processes contributing to sea level rise.

Past trends

In the past hundred years European and global average sea level has risen by 0.10 to 0.20 m with a central value of 0.15 m (IPCC, 2001a). Currently, the sea level at European coasts is rising at a rate of 0.8 mm/year (Brest and Newlyn) to 3.0 mm/year (Narvik) (Map 3.7, Figure 3.9), which is close to the global average trend.

A change in mean sea level at the coast may happen for various reasons. It can be caused by the vertical movement of the land itself (e.g. postglacial processes), by a local sea level change due to changes in the prevailing winds and ocean currents, or by a change in the volume of the world's oceans. It is very likely that the observed trend in sea level rise over the past 100 years is mainly attributable to an increase

in the volume of ocean water as a consequence of global climate change. The local variations could be explained by some of the other processes mentioned above.

Climate change related processes that contribute to the increase in ocean water volume are: the thermal expansion of water; the loss of mass of glaciers, ice caps and ice sheets; the runoff from thawing of permafrost; and

Source: D. Viner, 2004.

Figure 3.10 Projected global average sea level rise

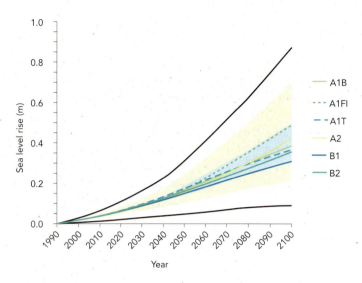

Note: Thermal expansion and land ice changes were calculated using a simple climate model calibrated separately for each of seven air/ocean global climate models (AOGCMs). Projections are for the six IPCC (SRES) scenarios. Light shading shows range of all models and scenarios. The dark lines include also the uncertanties in changes of land-ice, permafrost and sediment deposition. **Source:** IPCCa, 2001a.

the deposition of sediment on the ocean floor (IPCC, 2001a). The observed rise in sea level in the twentieth century tallies with the estimated rise from global climate models. This provides another indication that the observed sea level rise is, at least in part, caused by climate change (IPCC, 2001a).

Satellite data suggest a rate of sea level rise during the 1990s that is even greater than the mean rate of increase over much of the twentieth century (IPCC, 2001a; Nerem and Mitchum, 2001). The high thermal storage capacity of ocean water and the slow reaction of ice-shields appears to delay a general accelerated sea level rise at European coasts.

Projections (future trends)

Under the range of the six SRES scenarios (IPCC, 2001a) and a calibrated global climate/ocean model, a sea level rise of 0.09–0.88 metres has been projected for 1990 to 2100, with a central value of 0.48 cm (Figure 3.10). The central value equals an average rate of 2.2 to 4.4 times the rate over the twentieth century (IPCC, 2001a).

Even if greenhouse gas concentrations are stabilised, sea level will continue to rise for hundreds of years. After 500 years, sea level rise from the thermal expansion of oceans may have reached only half its eventual level, glacier retreat will continue and ice sheets will continue to react to climate change.

3.4.2 *Sea surface temperature*

Key messages

- Since the late nineteenth century, the global average sea surface temperature has increased by 0.6 ± 0.1 °C, consistent with the increase in global air temperature.
- Global ocean heat content has increased significantly since the late 1950s. More than half of the increase in heat content has occurred in the upper 300 metres of the ocean.
- No European sea shows a significant cooling; the Baltic and North Seas and the western Mediterranean show a slight warming of about 0.5 °C over the past 15 years.
- It is very likely that the oceans will warm less than the land; by 2100, global sea surface temperature is projected to increase by 1.1–4.6 °C from 1990 levels.

Figure 3.11 Annual sea surface temperature (SST) deviations averaged over the northern hemisphere

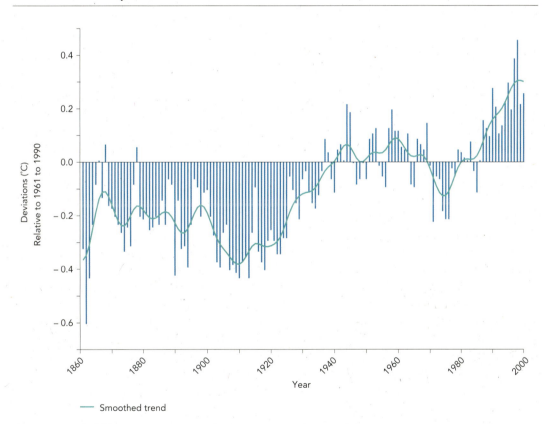

— Smoothed trend

Source: IPCC, 2001a.

Relevance

The oceans have a large capacity for storing and redistributing heat. By storing heat, they delay global temperature increase. Furthermore, oceans can attenuate the consequences of increasing CO_2 emission by absorbing atmospheric CO_2. However, oceans have been warming and, if this warming continues, the solubility of CO_2 and thus the uptake of CO_2 by the oceans will decrease. On the other hand, ocean warming can activate zoo- and phytoplankton, the so-called 'biological pump', which is responsible for the oceans' biological uptake of CO_2. However, the deteriorating physical solubility of CO$_2$ in warmer oceans would override the positive effect in terms of global average.

The changes in sea surface temperatures of global oceans are consistent with variations and changes in the global climate system, particularly the atmospheric temperature (see Section 3.2.2).

The consequences of increasing sea surface temperature include: rising sea level; reduced occurrence of Arctic sea ice; impacts on marine ecosystems, fisheries and aquaculture; and increasing risks for human health by enhanced epidemic bacteria and harmful algal blooms (see also the separate indicators on sea level rise, Arctic sea ice, marine growing season and marine species composition).

Measurement of sea surface temperature has been carried out by vessels, buoys and stationary equipment for many decades and, in recent decades, also by remote sensing. The uncertainty is low. For projections, remaining uncertainties are due to lack of measurement data and of knowledge of currents and their effects on sea surface temperature.

Past trends

Ocean temperatures are rising in all major ocean basins, with variations from ocean to ocean. This trend is consistent with the observed increase in air temperature (Levitus *et al.*, 2000; Cane *et al.*, 1997). Over the past 100 years, an initial warming phase (1910–1945) was followed by a period of cooling. A second warming phase began during the 1970s and is still continuing.

Figure 3.12 Sea surface temperature in winter and summer in the Norwegian Sea, the Baltic Sea and the western Mediterranean Sea

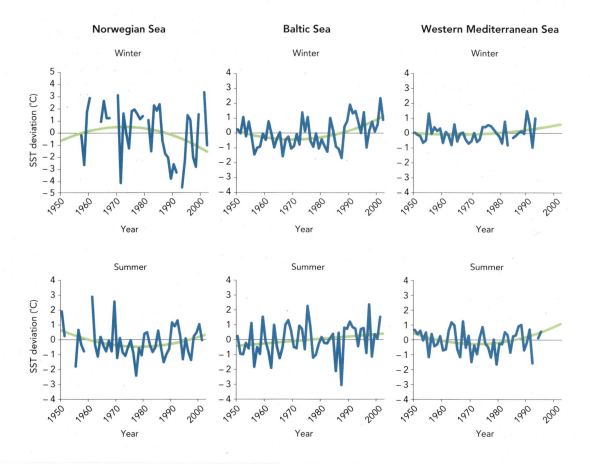

Note: Deviations are from long-term averages.
Source: Dooley, ICES, 2003.

The sea surface temperatures of the European seas are mainly influenced by regional weather patterns and circulation. Thus, the clear signal of warming in global sea surface temperature (Figure 3.11) is only partly reflected in the temperatures of the European seas (Figure 3.12).

Temperatures in the northern Seas, e.g. the Norwegian Sea, do not indicate a trend since they are strongly related to variations in regional atmospheric circulations (the North Atlantic Oscillation and the Arctic Oscillation) and to oceanic currents (e.g. strength and temperature of the warm North Atlantic Current) (Melsom, 2001; Furewik, 2000; Mizoguchi *et al.*, 1999). The North Sea and the Baltic Sea show a slight but not significant warming in winter and summer, particularly since the 1980s.

The western Mediterranean and the North Atlantic react in a similar manner to the global oceans. In the 1990s these seas warmed by about 0.5 °C (Rixen *et al.*, 2004).

Projections (future trends)

It is very likely that the oceans will warm less than the land since the greater heat capacity of the ocean

Algae bloom at the beach
Source: Van Liere, 2003.

delays warming (IPCC, 2001a). The projected increase of the global sea surface temperature by 2100 (relative to 1990) ranges from 1.1 ° to 4.6 °C (see Section 3.2.2).

Geographic variations in the amount of warming will occur, mainly caused by local and regional atmospheric and oceanic processes. Deep ocean mixing is likely to cause a limited warming of the surface of the North Atlantic and southern oceans. Warming in the North Atlantic may be further diminished by a weakening of the oceanic circulations (IPCC, 2001a).

3.4.3 Marine growing season

Key messages

- Increasing phytoplankton biomass and an extension of the seasonal growth period have been observed in the North Sea and the North Atlantic over the past decades.
- In the 1990s, the seasonal development of decapods larvae (zooplankton) occurred much earlier (by 4–5 weeks), compared with the long-term mean.

Figure 3.13 Long-term monthly means of phytoplankton colour index in the central North Sea

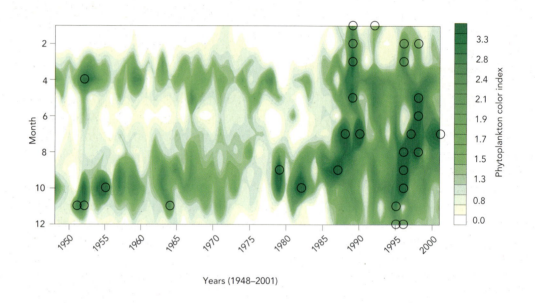

Source: M. Edwards, (SAHFOS), 2003.

Figure 3.14 Deviations of winter and summer sea surface temperature in the North Sea

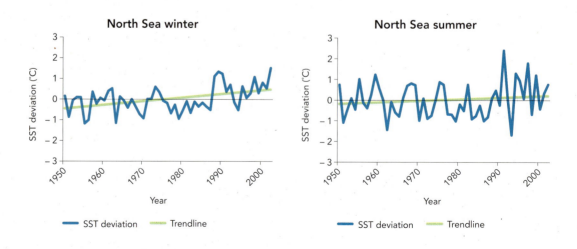

Source: H. Dooley, ICES, 2003.

Relevance

Global warming and the resulting warming of the oceans (see Section 3.4.2) can alter the marine growing season significantly. Variations in marine growing seasons affect mainly the development and biomass production of plankton.

Plankton, minute marine life that drifts with the currents, is the basis of the entire marine food web. Furthermore, plankton absorbs CO_2 from the upper ocean layers and moves it, together with nutrients, to the deep ocean.

This process, termed the 'biological pump', reduces the carbon content of the surface layers and maintains the ocean as a sink for atmospheric carbon.

Increased plankton production in coastal waters may be caused by enhanced nutrient inputs from human activities (e.g. waste water treatment plants, industry, households, agricultural fertilisers and traffic). However, it is likely that changes in plankton production and seasonality in the open sea are also driven by climate change (increase in sea surface temperature, changes in vertical mixing and cloud cover, which affects light and nutrient

The 'Continuous Plankton Recorder'
Source: SAHFOS, 2003.

supply). Furthermore, variations in the atmospheric nitrogen deposition affect plankton production, particularly in shallow seas.

An earlier start to plankton production in the season can enhance and change overall marine biological production, including a change in fish population, which may have an effect on fisheries production. Some phytoplankton species are toxic to higher forms of life (zooplankton, shellfish, fish, birds, marine mammals and, through the food chain, even humans) and may cause harmful algal blooms.

Data on plankton in the European seas are available for each year since about 1930 based on a huge amount of samples. Uncertainties result from the complexity of the indicator, which is sensitive to nutrient availability and influenced by many other factors such as the stability of the surface layer and the waterbody stratification as well as the radiation conditions.

Past trends

Phytoplankton biomass can act as an indicator for the marine growing season and has increased considerably over the past few decades in parts of the northeast Atlantic and the North Sea

Figure 3.15   Changes in the seasonal timing of decapod larvae in the North Sea

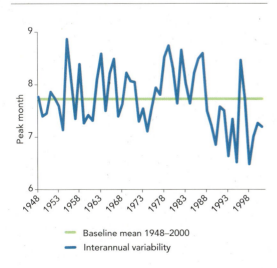

Baseline mean 1948–2000
Interannual variability

Source: Edwards, SAHFOS, 2002.

(Figure 3.13). In the late 1940s to the 1980s, the majority of production was restricted to bloom periods in spring and autumn. However, production has significantly increased during the winter and, especially, the summer season since the late 1980s (SAHFOS, UK, 2002). Particularly high increases have been observed since the mid-1980s in the North Sea and west of Ireland between 52 °N and 58 °N (Reid et al., 1998; Edwards et al., 2001).

During the 1990s phytoplankton biomass increased in winter months by 97 % compared with the long-term mean. Changes in annual phytoplankton biomass and the extension of the seasonal growing period (Figure 3.15) already appear to have considerable impacts on overall biological production and the food web.

The changes in sea surface temperature in the North Sea (see Figure 3.13 and Section 3.4.2) show similar spatial and temporal patterns to those of the phytoplankton biomass, indicating a relationship between sea surface temperature and length of growing season. The change in the seasonal timing of decapod larvae (as an example for zooplankton) over the period 1948–2000 shows a similar behaviour (Figure 3.15). Although there is considerable inter-annual variability of decapod larvae in the period 1948–2000, since 1988 the seasonal development of the larvae has occurred much earlier than the long-term average. The seasonal cycle was up to 4–5 weeks earlier in the 1990s than the long-term mean (Edwards et al., submitted).

Projections (future trends)

Increasing temperatures will cause further changes in the marine growing season and the productive capacity of marine and freshwater bodies. This will significantly change the productivity of marine ecosystems and also affect commercial fisheries.

3.4.4 Marine species composition

Key messages
- Over the past 30 years there has been a northward shift of zooplankton species by up to 1 000 km and a major reorganisation of plankton ecosystems.
- The presence and number of warm-temperate species have been increasing in the North Sea over the past decades.

Figure 3.16 Changes in species composition between a cold temperate and a warm temperate species of copepod in the North Sea

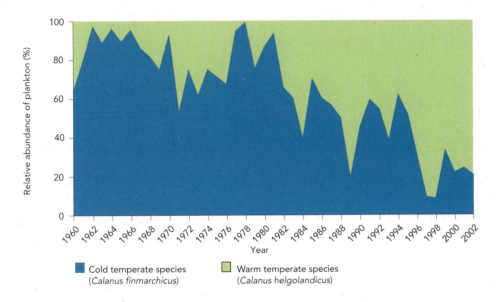

Cold temperate species
(*Calanus finmarchicus*)

Warm temperate species
(*Calanus helgolandicus*)

Source: Edwards, SAHFOS, 2003.

Relevance

The seasonal and inter-annual composition and the number of individuals per species of marine life depend strongly on the physical and chemical state of the marine environment. Climate change, particularly changes in northern hemisphere temperatures and the North Atlantic Oscillation (Beaugrand *et al.*, 2002), affects this state and therefore species composition.

Climate change can reduce the stability of marine ecosystems, increase the risk of disease, and may cause a loss of biodiversity. Changes in species composition also affect fisheries production.

Changes in marine biodiversity are addressed by the Conventions on

Biological Diversity and the Law of the Sea, regional conventions such as those of the Council of Europe, and the directives of the European Union.

Data on marine species composition are more certain than on phytoplankton concentration, due to the much lower sensitivity to other factors.

Past trends

Climate change related modifications in spatial distribution and phenology have been observed for many species in terrestrial ecosystems in Europe (see Section 3.5). These changes have also been detected in marine ecosystems. Substantial changes in marine species composition have been reported for the period from 1960 to 2002. Some zooplankton species have shown a northward shift of up to 1 000 km, in

Plankton-feeding basking shark
Source: D. Sims, MBA, 2003.

combination with a major reorganisation of marine ecosystems. These shifts have taken place southwest of the British Isles since the early 1980s and, from the mid 1980s, in the North Sea (Beaugrand *et al.*, 2002).

Most of the warm-temperate and temperate species have migrated northward by about 250 km per decade, which is much faster than the migration rate in terrestrial ecosystems (Parmesan and Yohe, 2002). In contrast, the diversity of colder temperate, sub-Arctic and Arctic species has decreased in this area (Beaugrand *et al.*, 2002). Furthermore, a northward extension of the ranges of many warm-water fish

species in the same region has occurred, indicating a shift of marine ecosystems towards a warmer northeastern Atlantic (Reid and Edwards, 2001).

Over the past decades there has been an invasion of warm-water/sub-tropical species into the more temperate areas of the northeast Atlantic. A useful indicator for this trend in the North Sea is the percentage ratio of two zooplankton species: the cold-temperate Calanus finmarchicus and the warm-temperate Calanus helgolandicus copepod (Figure 3.16). The results show a clear increase in the fraction of warm-temperate C. helgolandicus species, a trend that has accelerated over the past decade (Edwards *et al.*, 2002).

Projections (future trends)

Projected global warming in the twenty-first century is likely to have an effect on biological processes and biodiversity in the ocean, particularly in high latitudes (IPCC, 2001b). It is likely to cause more changes in the structure of biological communities in the upper oceans. Increasing temperatures will cause further changes in the productive capacity of marine and freshwater bodies, and will trigger a shift in species distribution and an increase in diversity towards higher latitudes (v. Westernhagen *et al.*, 2001). More warm-water species will migrate northwards and compete for existing niches (IPCC, 2001c).

3.5 Terrestrial ecosystems and biodiversity

3.5.1 Plant species composition

Key messages
- Climate change over the past three decades has resulted in decreases in populations of plant species in various parts of Europe.
- Plant species diversity has increased in northwestern Europe due to a northward movement of southern thermophilic species, whereas the effect on cold tolerant species is still limited.
- Projections predict a further northward movement of many plant species. By 2050, species distribution is projected to become substantially affected in many parts of Europe.
- Globally a large number of species might become extinct under future climate change. Due to non-climate related factors, such as the fragmentation of habitats, extinction rates are likely to increase. These factors will limit the migration and adaptation capabilities needed by species to respond to climate change.

Source: www.bigfoto.com, 2004.

Figure 3.17 Changes in frequencies of groups of plant species adapted to 'warm' and 'cold' conditions in the Netherlands and Norway

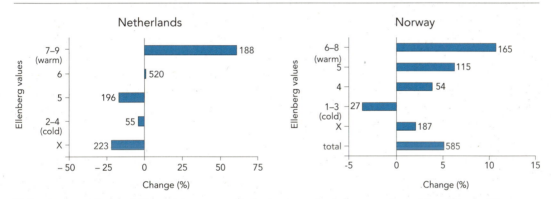

Note: Low numbers on the y-axis represent plant groups that are accustomed to cold conditions, high numbers are groups accustomed to warm growing conditions (according to Ellenberg numbers). In the Netherlands, the periods 1975–1984 and 1985–1999 are compared; for Norway, the periods 1958–1961 and 2000–2002 are compared.
Source: Tamis et al., 2001; Often and Stabbetorp, 2003.

Relevance

Plant species can successfully reproduce and grow only within specific ranges of climatic conditions. If these conditions change, they either have to adapt or to migrate. For some species, particularly in high elevation and northern areas, migration is often difficult. If both types of responses are not feasible, local populations of species will become extinct. Changes in plant species composition in a given area are significantly influenced by climate change, but other factors such as land-use changes also play an important role.

In turn, decreasing plant species richness limits overall biodiversity, which can lead to a decrease in ecosystem stability and threaten certain ecosystem goods and services (e.g. in relation to the production of medicines). Furthermore, changes in plant species distribution and thus the regional composition of vegetation may have consequences on the climate system. In high latitudes, for example, the replacement of shrubby tundra vegetation by trees can have a considerable effect on the radiation balance (especially in cases of snow cover), which in turn is likely to enhance regional and global climate change.

The extent to which changes in plant species composition will occur may be limited by policy responses (IPCC, 2001b; CBD, 2003). The establishment of a network of protected areas (such as Natura 2000 in Europe), for example, could prevent some decline in species richness if these areas are successfully managed.

Uncertainty over the response of plant species to climate change is modest. There is still a lack of accurate data on the effect of climate change on plant species diversity across Europe. Some recent projects in the EU may provide new data in the near future (e.g. the UK Monarch project; http://www.eci. ox.ac.uk/biodiversity/monarch.html; or the ATEAM project; http://www.pik-potsdam.de/ateam/).

Past trends

In many parts of the world, including Europe, species composition has changed and species have become extinct at rates 100–1 000 times greater than is considered to be normal (IPCC, 2002; Hare, 2003). Although most changes are attributable to landscape fragmentation and habitat destruction, studies show a high correlation between changes in plant composition and recent climate change (e.g. Hughes, 2000; Pauli *et al.*, 2001; Parmesan and Yohe, 2003). This high correlation is based on the fact that climate ultimately determines the distribution of plant species, the frequency of natural disturbances like forest fires (e.g. in southern Europe and Russia) and nutrient availability due to changes in soil composition.

Map 3.8   Share of stable species in 2100, compared with 1990

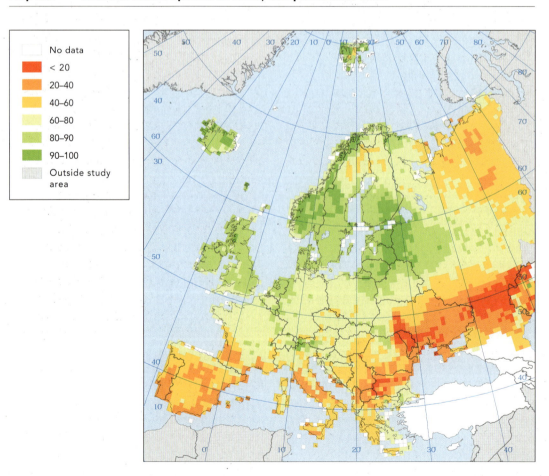

Note: Percentage of total number of species in 1990. The climate scenario used is a modest climate change scenario (global warming by 2100 is 3 °C and European warming is 3.3 °C).
Source: Bakkenes *et al.*, 2004.

Over the past decades a northward extension of various plant species has been observed in Europe which is likely to be attributable to increases in temperatures (CBD, 2003; Parmesan and Yohe, 2003). Many Arctic and tundra communities are affected and have been replaced by trees and dwarf shrubs (Molau and Alatalo, 1998). In northwestern Europe, for example in the Netherlands (Tamis *et al.*, 2001), the United Kingdom (Preston *et al.*, 2002), and central Norway (Often and Stabbetorp, 2003), thermophilic (warmth demanding) plant species have become significantly more frequent compared with 30 years ago (in the Netherlands, by around 60 %). In contrast, there has been a small decline in the presence of traditionally cold-tolerant species (Figure 3.17). The changes in composition are the result of the migration of thermophilic species into these new areas, but also due to an increased abundance of these species in current locations.

Projections (future trends)

The impact of climate change on plant species composition will increase in the coming decades. Future climate change is estimated to exacerbate the loss of species, especially those species with restricted climate and habitat requirements and limited migration capabilities (IPCC, 2001b). A 3 °C increase in temperature, within the range projected for 2100, corresponds to a shift in species distribution of 300–400 km to the north (in the temperate zones) or 500 m in elevation (Hughes *et al.*, 2000). Many species will have difficulties in responding to such rapid change by migration or adaptation, and

Source: R. Müller, www.pixelquelle.de, 2004.

are likely to become more restricted in distribution or even extinct (Root *et al.*, 2003). Thomas *et al.* (2004) projected that, under such conditions, 15–37 % of all species globally might become extinct by 2050. In mid- and northern Europe, extinct plant species may be replaced by thermophilic species. The greatest effects are projected for Arctic regions, the moisture-limited ecosystems of eastern Europe and the Mediterranean region (Map 3.8) (Bakkenes *et al.*, 2004). Current plant species richness in the Mediterranean area might be reduced over the twenty-first century because of the projected decreases in precipitation, more frequent forest fires, increased soil erosion and the lack of species that could replace those that are lost. Endemic species in northern Europe may become extinct and replaced by more competitive species in the long term (e.g. Sykes and Prentice, 1996; CBD, 2003).

3.5.2 Plant species distribution in mountain regions

Key messages

- Endemic mountain plant species are threatened by the upward migration of more competitive sub-alpine shrubs and tree species, to some extent because of climate change.
- In the Alps, upward migration has led to an increase in plant species richness in 21 out of 30 summits, whereas it has decreased or remained stable in the other summits.
- Projected changes in European annual average temperature are outside the tolerance range of many mountain species. These species are projected to be replaced by more competitive shrub and tree species, leading to considerable loss of endemic species in mountain regions.

Figure 3.18 Change in species richness on 30 high summits of the eastern Alps during the twentieth century

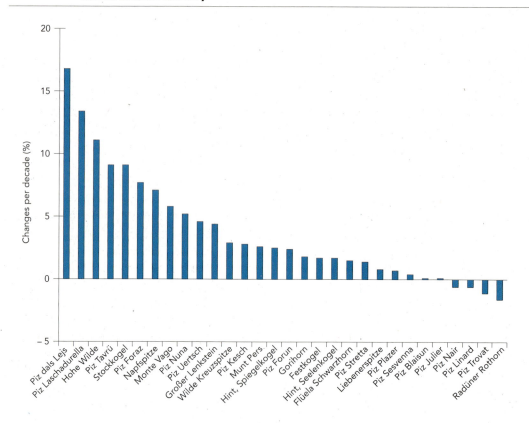

Source: Grabherr *et al.*, 2002.

Relevance

Mountain regions are important for the diversity of European flora. Europe's mountain zones above the tree line include approximately 20 % of all native European vascular plant species (Väre *et al.*, 2003). In these regions, climate is the main driver of species composition

and human influence is relatively low (with some exceptions). Mountain plant species are vulnerable to climate change because these species are: a) likely to respond more to enhanced CO_2 levels than low altitude vegetation; b) characterised by small climatic ranges, severe climatic conditions and small isolated populations (Pauli *et al.*,

2003); and c) unable to migrate because of the absence of suitable areas at higher elevations. Fortunately, mountain regions are also often characterised by many microscale regions with different climates, in which endemic species might survive even if the wider climate changes beyond tolerance limits (Körner, 1995).

Consequences of climate change are likely to include the extinction of rare endemic species in mountain regions, as well as changing the appeal of such regions (e.g. for tourism). The response of mountain plant species to climate change is still relatively uncertain. There is a lack of accurate data and of knowledge (e.g. about adaptation). Within the most comprehensive European study (GLORIA — Global observation research initiative in Alpine environments), monitoring sites have been in operation for only a few years. Further uncertainties in projecting future trends are due to absence of local-scale climate information.

Alpine plant Edelweiss
Source: Th. Fabbro, www.unibas.ch/botimage, 2004.

Past trends

Endemic species have been replaced by other species in the mountain regions of Europe due to a number of factors, including climate change. Higher temperatures and longer growing seasons, associated with climate change, have created suitable conditions for certain plant species that have migrated upward and which now compete with the endemic species (Grabherr *et al.*, 1994; Gottfried *et al.*, 1999). The net effect on species richness diverges from region to region and even within single regions. While richness has increased in some places, it has declined in others. In the Alps, for example, evidence exists that climate warming over the past 60 years has encouraged spruce and pine species in the sub-alpine region (Pauli *et al.*, 2001) and sub-alpine shrubs to grow on the summits (Mottas and Masarin, 1998; Theurillat and Guisan, 2001). The net effect is an increase in species richness in 21 out of 30 summits in the Alps compared with 50 to 100

years ago (Grabherr *et al.*, 2002, Figure 3.18). Similar trends have occurred in the Pyrenees, Scandinavia, Bulgaria and the Ural (Klanderud and Birks, 2003; Kullman, 2003; IPCC, 2001b; Meshinev *et al.*, 2000; Montserrat, 1992).

Projections (future trends)

Future climate change is projected to affect species distribution in mountain regions considerably, resulting in decreased abundance and sometimes even losses of endemic species. These species might become threatened as they will be unable to adapt to the changed environment, cannot migrate to better places and cannot compete with immigrating shrub and tree species (IPCC, 2001b; Pauli *et al.*, 2003). In the lower Alps, the tree line will climb and competition from Norway spruce will likely cause the deterioration of growing conditions for endemic plant species (Theurillat and Guisan, 2001). For Scandinavia, it is projected that there will be a 40–60 % reduction of the current mountain vegetation area (Holten and Carey, 1992).

Bakkenes *et al.* (2002) has projected considerable effects on selected species in many mountain regions in Europe, even under a modest climate change

Map 3.9   The potential response by 2100 of three currently common mountain species to climate change

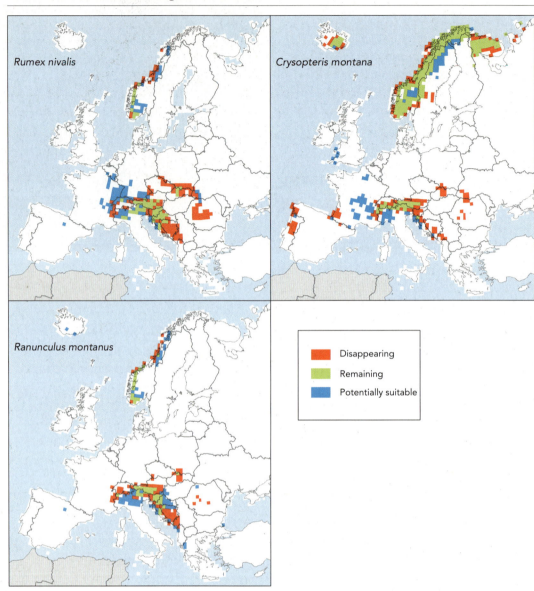

Note: Snow dock (*Rumex nivalis*), mountain bladder fern (*Cystopteris montana*), and mountain buttercup (*Ranunculus montanus*). Projections are based on using the EuroMove model under a modest climate change scenario (global warming by 2100 is 3.0 °C and European warming is 3.3 °C, respectively).
Source: Bakkenes *et al.*, 2004.

scenario in the range of IPCC scenarios (Map 3.9). For example, mountain bladder fern (Cystopteris montana) will disappear by 2100 in 20–30 % of its current locations (Map 3.9).

There are various uncertainties in projecting the impacts of climate change on species composition in mountain regions of Europe. Firstly, the accuracy of climate projections is limited in simulating small-scale plant communities. Secondly, mountain plant species are only modestly represented in the simulation models. Thirdly, the migration rates of many tree species might not be as great as projected since the development of appropriate growing conditions comprises more than climate and could require decades. This may limit the projected extinction of species. Finally, the adaptive capacity of mountain plant species is often unclear and might change in the future (see also Chapter 5).

3.5.3 Terrestrial carbon uptake

Key messages

- In the period 1990–1998 the European terrestrial biosphere was a net sink for carbon and therefore partly offset increasing anthropogenic CO_2 emissions.
- Carbon uptake in Europe can be increased by (re-)planting forests and other land management measures. The additional potential storage capacity for the EU in forestry and agriculture is estimated to be relatively small, compared with the agreed targets in the Kyoto Protocol.
- The projected increase in average temperature is likely to reduce the potential amount of carbon that can be sequestrated in the European terrestrial biosphere in the future.

Map 3.10   Annual carbon uptake of the terrestrial biosphere

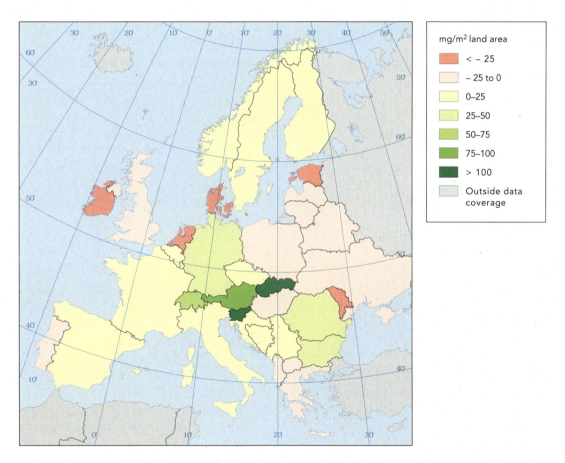

Note: The biosphere of countries with positive values absorbs more carbon than it emits (carbon sink); those with negative values emit more than they absorb (carbon source).
Source: Janssens *et al.*, 2003, CarboEurope.

Relevance

The uptake of carbon by natural vegetation, soils, forests and agricultural land ('terrestrial biosphere') is an important part of the carbon cycle. Terrestrial carbon uptake can lessen the increasing concentration of CO_2 in the atmosphere. This increase is mostly driven by anthropogenic emissions. The uptake rate and storage capacity of carbon in the terrestrial biosphere are influenced by both natural factors and human activities, including temperature, precipitation, atmospheric CO_2 concentration, nitrogen fertilisation by air pollutants (nitrogen oxides and ammonia), the growth rate of plants, fire, storms, land-use changes, pests and the harvesting and trading of agricultural and forestry products.

In recent decades the European terrestrial biosphere has been a net sink of carbon, but with large differences between countries (Map 3.10). Active land-use management mainly in forests (e.g. by replanting) but also in agriculture (e.g. by raising the carbon content of soils) could further increase recent carbon sequestration. The projected rise in carbon storage by active land management between 1990 and 2008–2012 will help to achieve the Kyoto Protocol targets. According to the UNFCCC Marrakech Accords, these responses include afforestation, reforestation and other land use and forestry activities (re-vegetation, forest management, cropland management and grazing land management; see Kyoto Protocol Art 3.3 and 3.4; see also Section 2.2)

There are still data gaps and a lack of knowledge in understanding the processes of carbon sequestration. Hence, uncertainty regarding the total size of current and future European carbon sequestration is high.

Carbon flux measuring tower in a European forest
Source: M. Schumacher, CarboEurope, 2002.

Past trends

The recent European carbon uptake from the atmosphere into the biosphere is very low, less than 1.5 % of the total carbon exchange between the biosphere and the atmosphere (Figure 3.19, IPCC 2001a). The terrestrial carbon cycle is strongly linked to inter-annual climate variability (Figure 3.20). However, the rising atmospheric CO_2 concentration, nitrogen fertilisation, reduced air pollution and changed management have resulted in a steady increase in annual forest storage capacity in the past few decades, which leads to a more significant carbon uptake (Nabuurs *et al.*, 2002).

During the 1990s the European terrestrial biosphere stored between 7 % and 12 % of the annual anthropogenic CO_2 emissions (Janssens *et al.*, 2003). Carbon budget per country (Map 3.10) shows that about half of the European countries are significant carbon sinks (Austria, Bulgaria, Germany, Norway, Romania, Slovakia, Slovenia, Sweden, Switzerl and) or sources (Denmark, Estonia, Ireland, Moldova, the Netherlands). The other countries are weak sources or sinks. However, uncertainties in the calculations of the national and European carbon flux density are still high (Figure 3.20, blue and yellow lines).

Projections (future trends)

The projected increase in temperature and atmospheric CO_2 concentration will most likely improve growing conditions in mid and northern Europe and enhance carbon storage, especially in boreal forests. The additional potential carbon storage capacity for the EU by 2008–2012, through various measures in forestry and agriculture, might be equivalent to about 2.4 % of the anthropogenic greenhouse gas emissions, which would be about 30 % of the reduction target agreed in the Kyoto Protocol. The accountable sink capacity according to the Kyoto Protocol (Art. 3.3 and 3.4) is much lower, namely about 1 % of anthropogenic GHG

Figure 3.19 Carbon balance of the terrestrial biosphere

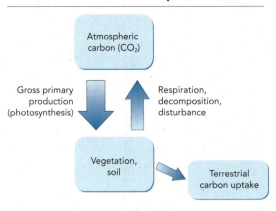

Notes: Atmospheric carbon (in the form of CO_2) is absorbed by plants in the process of photosynthesis. A large part of this carbon is released back to the atmosphere. A small part is removed from the atmosphere and stored in the terrestrial biosphere (terrestrial carbon uptake).
Source: M. Zebisch, 2004.

emissions, equivalent to about 12 % of the total reduction target (EEA, 2004).

However, there are several processes which could reduce carbon sequestration in future. Climate change may cause more fires, pests and storm damage as well as increasing water stress, particularly in the Mediterranean area. These conditions would curtail plant growth and reduce the amount of carbon stored in the biosphere. An increase in plant and soil respiration due to higher temperatures in future may further reduce carbon storage in many ecosystems (Rustad *et al.*, 2001). Human management measures to maximise carbon storage in terrestrial ecosystems to meet the Kyoto targets may be restricted by saturation effects in the biosphere. The potential storage capacity by forests is limited by the fact that only two-thirds of the annual increment is currently harvested (UN-ECE/FAO — UN Economic Commission for Europe/Food and Agriculture Organisation, 2000). Combining all these factors, the European biosphere's ability to sequester carbon may be reduced significantly in the future. The biosphere may even turn from a sink into a source of carbon. A better understanding of the processes is required to achieve more reliable projections of carbon sequestration in the twenty-first century.

Figure 3.20 Inter-annual variation in European carbon fluxes from the biosphere to the atmosphere

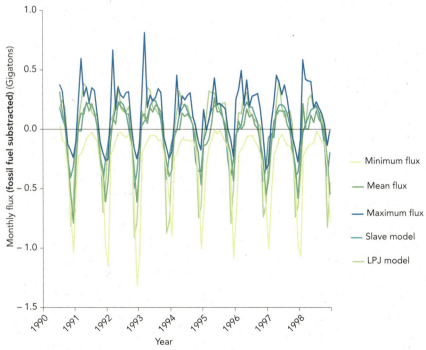

Note: The lines represent the results of two ecosystem models (LPJ and SLAVE), and the range (minimum, mean, maximum) of calculated terrestrial C flux (based on so-called inversed calculations considering atmospheric CO_2 calculation). The biosphere is a sink for carbon if values are negative.
Source: Bousquet *et al.*, 2000.

3.5.4 Plant phenology and growing season

Key messages

- The average annual growing season in Europe lengthened by about 10 days between 1962 and 1995, and is projected to increase further in the future.
- Greenness (a measure of plant productivity) of vegetation increased by 12 %, an indicator of enhanced plant growth.
- The positive effects of temperature increase on vegetation growth (i.e. a longer growing season) are projected to be counteracted by an increased risk of water shortage in mid and especially southern Europe which would negatively affect vegetation.

Figure 3.21 Observed changes in growing season length

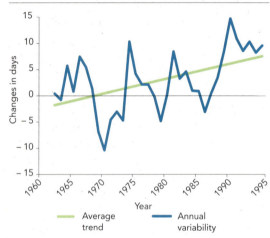

Source: M. Zebisch, 2004.

Note: Observed data from the International Phenological Gardens in Europe except France, the Iberian peninsula, mid and southern Italy, and Greece.
Source: Menzel, 2002; Menzel u. Fabian, 1999.

Relevance

Plant species are adapted to a specific range of climate and location conditions. Their presence is therefore restricted to a distinct geographical area. Plant growth is determined by temperature, precipitation and atmospheric CO_2 concentration. In response to a warming climate, some species are able to grow better than others. Indigenous species might be replaced by new species which are better adapted to higher temperatures and/or increasing drought stress. Some crops and trees need low temperatures in winter to trigger bud bursting in spring. These species can be adversely affected in areas where winter temperatures become too high (Chuine and Beaubien, 2001).

As climate change may lengthen the growing season, the period between bud burst and leaf fall is projected to extend further, particularly in mid and northern Europe. This will lead to an increase in biomass production in areas where temperature has been the limiting factor of plant growth. On the other hand, warming will increase the risk of drought stress in the lower altitudes of mid and especially southern Europe, where lack of water is already a limiting factor.

Changes in growing season can also affect species composition, in particular those with low adaptive capacity (see Section 3.5.1). Economically important tree species may no longer be suitable for forestry and may have to be replaced (Sykes *et al.*, 1996; Parry 2000). Consequently, changes in growing season due to climate change will lead to changes in agricultural and forest management. Nature protection will be affected too if the growth and survival of protected plant species are endangered by climate change (see Section 3.5.1).

Growing season change is closely related to climate change and has a relatively low uncertainty. However, gaps in knowledge and data exist about the impact of a changing growing season on plant growth and on biodiversity.

Figure 3.22 Greenness of vegetation in Europe

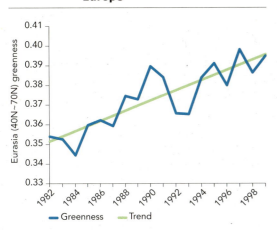

Note: Greenness is a measure for plant productivity derived from remote sensing data.
Source: Zhou *et al.*, 2001.

Past trends

The length of the growing season is very sensitive to climate. Phenological data (Figure 3.21) show a clear increase in the length of the growing season by about 10 days from 1962 to 1995 (Menzel and Fabian, 1999). The overall trend towards longer growing seasons is consistent with an increase of about 12 % in the 'greenness' of vegetation (Figure 3.22, Zhou *et al.*, 2001). Greener biomass (needles and leaves) indicates an increase in plant growth. The extension of the growing season in both spring and autumn, mostly coupled with higher temperatures during the growing period, appears to enhance the productivity

of vegetation in Europe. Plants that grow better under these conditions are likely to benefit from climate change. However, plants which cannot cope with the increasing temperatures might be unable to compete and may be replaced by other species. Regional trends can vary from the European averages; for example, the growing season has shortened in the Balkan region (Menzel and Fabian, 1999).

Projections (future trends)

Climate change scenarios indicate a further increase in the length of the growing season (Figure 3.23) and in drought stress (Figure 3.24). Drought stress increases when the water demand of the plants exceeds water availability.

Biomass production in the boreal region (Scandinavia and northern Russia) is very likely to benefit most from increasing temperatures, with a 30 % longer growing season over the next 100 years, combined with almost no drought stress. Temperate vegetation may benefit from a 20 % longer growing season but suffer from a slight decrease (around 4 %) in water availability (higher drought stress). In southern Europe, the increasing risk of drought stress (+ 13 %) combined with only an 8 % longer growing season will likely affect vegetation growth negatively, mainly in areas of low elevation.

Figure 3.23 Projected no. of growing days

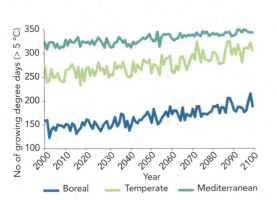

Note: Growing season is defined here as the number of consecutive days per year with average temperatures of more than 5 °C.
Source: LPJ model (Sitch *et al.*, 2003)

Figure 3.24 Projected drought stress

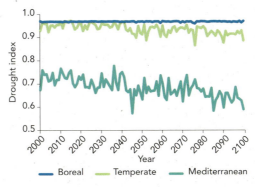

Note: Higher numbers indicate lower drought stress.
Source: LPJ model (Sitch *et al.*, 2003). Climate data from Mitchell *et al.*, 2004.

3.5.5 Bird survival

Key messages

- The survival rate of different bird species wintering in Europe has increased over the past few decades.
- The survival rate of most bird species is likely to improve further because of the projected rise in winter temperature.
- Nevertheless, it is not yet possible to determine what impact this increasing survival will have on bird populations.

Figure 3.25 Survival of grey heron and common buzzard

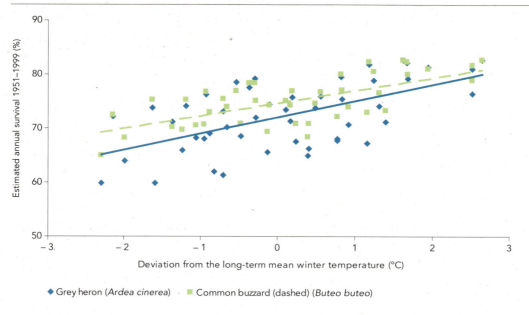

♦ Grey heron (*Ardea cinerea*)   ■ Common buzzard (dashed) (*Buteo buteo*)

Source: Frederiksen, 2002.

Relevance

The population demography of birds depends on both survival (proportion of birds that survive from year to year) and productivity (number of birds produced per year). Changes in bird populations will affect biodiversity and ecosystem functions. This issue is important in the context of the implementation of the EU biodiversity strategy and the EU birds directive.

Thus, survival is one of two key elements that determine the population dynamics of birds. For various bird species wintering in Europe, there is a correlation between winter temperature and survival, very likely because foraging is made easier when neither frost nor snow cover appears. The effect on bird populations is complicated

to predict, because the populations' adaptation to increased survival and climate change is very flexible and species interactions (predator/prey relations, competition for resources) will also be affected. Some species will benefit and their population will increase while others could be adversely affected.

The uncertainty in the observed trend is great, although a clear correlation has been found. But datasets are rare, relationships are complex and various non-climate factors play a role.

Past trends

In the past, the survival of some European bird species wintering in Europe increased between 2 % and 6 % per 1 °C rise in winter temperature, depending on species (see Figure 3.25) (Frederiksen, 2002). Since the increase in — particularly — winter temperatures due to climate change (see Section 3.2.2), this effect has been considerable. The correlation between bird survival and winter temperature has been observed for the grey heron, common buzzard, cormorant, song thrush and the redwing.

The higher survival rate clearly affects population demography, but the effect on population size is less obvious because of other factors determining population dynamics (e.g. productivity). Certain elements of bird biology (e.g. egg laying dates, migration dates) are known to vary according to climate change. Some of these factors are strongly dependent on temperature.

Projections (future trends)

Models have been used to determine future survival based on projected temperature increase in regions in Europe. These projections show a continuation of the trend presented in Figure 3.25 under the assumption that other parameters remain unchanged. At species level, the projections show that the effect of a better survival rate appears to be more predominant for

Common buzzard
Source: G. Whitlow.

populations in northern Europe than in other parts. The song thrush population in the UK, for example, responds significantly less to changing winter temperature than does the Finnish population. Assuming a modest climate change scenario (Parry, 2000), the projected temperature rise of 1.4–1.8 °C by 2080 in the UK could increase the survival rate of the song thrush by about 5 %, whereas the rate would improve by 12 % in Finland (with a temperature increase of about 1.8 °C).

3.6   Water

3.6.1   Annual river discharge

Key messages

- Annual river discharge has changed over the past few decades across Europe. In some regions, including eastern Europe, it has increased, while it has fallen in others, including southern Europe. Some of these changes can be attributed to observed changes in precipitation.
- The combined effect of projected changes in precipitation and temperature will in most cases amplify the changes in annual river discharge.
- Annual discharge is projected to decline strongly in southern and southeastern Europe but to increase in almost all parts of northern and northeastern Europe, with consequences for water availability.

Map 3.11   Changes of the mean annual river discharges over the twentieth century

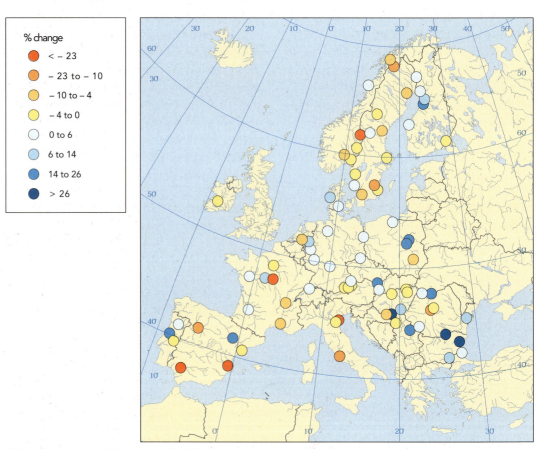

Note: The observed time periods differ between stations.
Source: CEDEX, Spain, 2003, based on UNESCO, 1999.

Relevance

The annual discharge of rivers, a measure of the stream flow, can be used as an indicator of climate change since it represents the response of an entire river basin to meteorological factors such as precipitation or temperature.

The annual river discharge is an indicator for fresh water availability in a river basin, and also a first estimate for low and high river flow events. If the annual river discharge increases, the risk of floods rises. Low annual river discharge could lead to more frequent low flow events, which would affect river transportation possibilities, for instance.

The most important policy framework in Europe related to river discharge is the water framework directive. According to the directive, each EU Member State has to establish a monitoring programme for water flows and water volumes in their major river basins. With the help of these data, observed river discharges could be related to climate data, and projections of future trends due to climate change could be made.

The uncertainties in the measurements of annual river discharge are relatively low (Kaspar, 2004) because data availability is good and processes are well understood. However, detecting whether and how climate change had an influence is complex since river flow depends on changes in both precipitation and temperature. The uncertainty in projecting future trends is considerable due to uncertainties in future global and especially regional climate.

Past trends

Annual river discharge varies widely across Europe, reflecting the continent's varied climate and topography. Reflecting patterns of precipitation, discharge levels range from very high in western Norway, on Britain's west coast and in southern Iceland to low in parts of Spain, Sicily, the Ukraine and Turkey. In some arid regions, the discharge volume is determined by precipitation in the upper parts of a basin. Hungary, for example, receives most of its water from outside the country borders through the Danube River.

The observed trends in discharge over the twentieth century show different patterns in river basins across Europe

Rhine at Schaffhausen
Source: www.bigfoto.com, 2004.

(Map 3.11). River discharge decreased considerably in many southern European basins such as the rivers Jucar and Guadalquivir (both in Spain), the Loire (France) and the Adige (Italy). In contrast, large increases in discharge occurred in eastern Europe (e.g. along the Danube). In central Europe, only small changes in annual river discharge occurred (e.g. the Rhine). Fresh water input to the Baltic Sea did not change between 1920 and 1990 (Winsor, 2001).

It is very likely that these changes are largely due to precipitation changes, although discharge has also been affected by various other factors such as land-use change or the straightening of rivers for shipping.

Projections (future trends)

Changes in annual river discharge are projected to vary significantly across Europe, related to regional/local changes in precipitation and temperature (see Section 3.2). By 2070, river discharge is expected to decrease by up to 50 % in southern and southeastern Europe, and to increase by up to 50 % or more in most parts of northern and northeastern Europe (Map 3.12). As a result, stress on water resources may continue to grow significantly in southern Europe.

Projected percentage changes in annual river discharge are often more pronounced than changes in

Map 3.12   Change in annual average river discharge for European river basins in the 2070s compared with 2000

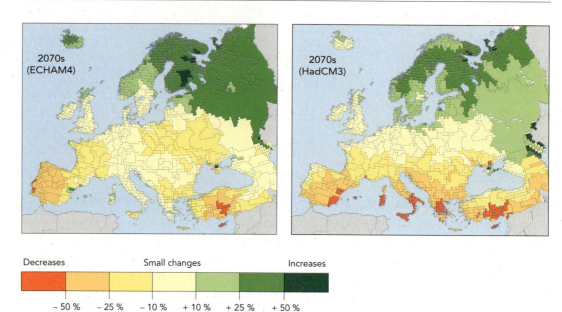

Decreases   Small changes   Increases

− 50 %   − 25 %   − 10 %   + 10 %   + 25 %   + 50 %

Note: Two different climate models (ECHAM4 and HadCM3).
Source: Lehner *et al.*, 2001.

precipitation, because precipitation only partly results in discharge (the other part mainly evaporates). Furthermore, in saturated soils, all additional precipitation will lead to discharge.

Climate models show large differences in their projections of regional changes in precipitation (Map 3.12). The uncertainty of projected river discharge is therefore also high.

3.7 Agriculture

3.7.1 Crop yield

Key messages

- The yields per hectare of all cash crops have continuously increased in Europe in the past 40 years due to technological progress, while climate change has had a minor influence.
- Agriculture in most parts of Europe, particularly in mid and northern Europe, is expected to potentially benefit from increasing CO_2 concentrations and rising temperatures.
- The cultivated area could be expanded northwards.
- In some parts of southern Europe, agriculture may be threatened by climate change due to increased water stress.
- During the heatwave in 2003, many southern European countries suffered drops in yield of up to 30 %, while some northern European countries profited from higher temperatures and lower rainfall.
- Bad harvests could become more common due to an increase in the frequency of extreme weather events (droughts, floods, storms, hail), and pests and diseases.

Map 3.13   Wheat yield in 2003 (change from 2002)

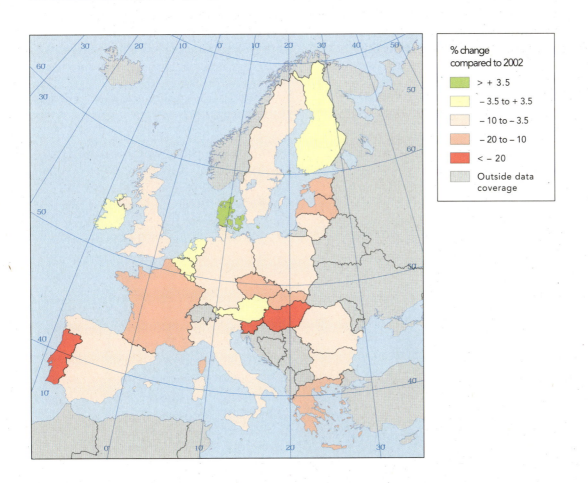

% change compared to 2002

- > + 3.5
- − 3.5 to + 3.5
- − 10 to − 3.5
- − 20 to − 10
- < − 20
- Outside data coverage

Source: Joint research centre (JRC) MARS project (Monitoring agriculture with remote sensing unit), 2003.

Relevance

Europe is one of the world's largest and most productive producers of food and fibre. The agricultural sector in Europe contributes 2.6 % to European GDP and employs 5 % of all workers. 44 % of the area is under agricultural use (European Commission, 2002).

Agriculture is very sensitive to climate change since climate is one of the most important natural factors for agricultural production. Rising CO_2 concentrations can stimulate photosynthesis, increase biomass production and enhance water use efficiency (Pinter *et al.*, 1996; Kimball *et al.*, 1993). This affects most European cash crops, such as wheat, barley, rye, potato and rice. Other crops, like maize, profit less. Rising temperatures can have different effects. They can enhance plant productivity and decrease the risk of damage by freezing. However, under warm and dry conditions, they can lead to water stress and, consequently, to a decline in yield. Changes in the amount and distribution of rainfall can affect agriculture positively or negatively, depending on the regional conditions and the regional trend.

Other effects of climate change include: a higher risk of bad harvests resulting from an increasing frequency and intensity of weather extremes (droughts, floods); a climate related increase in pests and diseases; and a decrease in crop quality (lower content of protein and trace elements) due to rising CO_2 concentrations (IPCC, 2001b).

Source: D. Viner, 2004.

Much is known about how climate change affects crop growth in general. On the other hand, the contribution of climate change to the observed trend in crop yield is subject to high uncertainties due to complex interactions between technical progress, policy measures and crop yield in the EU.

Past trends

The yields per hectare of all cash crops have increased worldwide in the last 40 years. This trend can mainly be explained by technological success in breeding, pest and disease control, fertilisation and mechanisation (Hafner, 2003). Besides technical progress, agriculture in Europe is largely determined by the Common Agricultural Policy (CAP). It is also affected by the cut in and decoupling of subsidies for agricultural products since the EU agricultural reform of 1993, which has already led to a reduction in the area under cultivation. The attribution of climate change to these trends is likely to be small. One positive effect of climate change that is already apparent is the earlier sowing date for certain crops due to an earlier start and longer duration of the vegetation period (see also Section 3.5.4).

In the past decades weather extremes have adversely influenced yield results. One of the most remarkable was the heatwave in 2003. High temperatures and a long period with low or no precipitation led to droughts in large parts of Europe. The consequent drop in crop yields was the strongest negative deviation from the long-term trend in Europe in the last 43 years (FAO, 2004). While most countries, including Greece, Portugal, France, Italy and Austria, suffered from yield drops of up to 30 %, some countries (Denmark, Finland) profited from the higher temperatures and lower rainfall (Map 3.13). The total harvest in the EU of most cash crops dropped substantially. The consequences of the 2003 heatwave are especially relevant because the extreme situation in 2003 is an example of what could be the average climate in the long-term future (2071–2100) (Beniston, 2004).

Projections (future trends)

Climate change is expected to boost yields for most crops in most parts of Europe over the coming decades. The magnitude of this effect is still uncertain and depends on the climate scenario and how agriculture adapts to climate change. Estimations show yield increases of 9 % to 35 % for wheat by 2050 (Hulme et al., 1999). The largest increases in yield could occur in southern Europe, particularly northern Spain, southern France, Italy and Greece. Relatively large yield increases (3–4 t/ha) may also occur in Scandinavia. In the rest of Europe, yields could be 1–3 t/ha greater than at present. There are small areas where yields are projected to decrease by as much as 3 t/ha, such as in southern Portugal, southern Spain, and the Ukraine (Harrison et al., 2003; IPCC, 2001). A critical factor in these scenarios is water supply and the uncertainty in projections of regional precipitation. As happened during the heatwave in 2003 (Map 3.13), a lack of precipitation could convert the positive effect of climate change (stimulated plant growth) into a negative effect (decrease in yield due to water stress) (Olesen and Bindi, 2002). This threatens particularly the southern and eastern parts of Europe (Spain, Greece).

In the long term, the area suitable for agriculture will likely shift northwards. In Finland, for instance, the agricultural area could expand northwards by 100–150 km per 1 °C temperature rise (Carter and Saarikko, 1996) while agricultural areas in the drier Mediterranean part of Europe may be abandoned. The sowing date for many crops could be brought forward, e.g. five to 25 days earlier for wheat (Harrison et al., 2003). On the other hand, any direct yield gain caused by increased CO_2 could be partly offset by losses due to changes in the spatial distribution and intensity of pests and diseases (IPCC, 2001).

The complex interaction of various factors means that the impact of climate change on agriculture in future is subject to many uncertainties. It will, to a great extent, depend on how agriculture can adapt to the expected climate changes. In the past, agriculture has shown a high adaptive capacity to pressures like growing population and declining prices. In future, climate change will be one of many pressures, such as the increasing rate of soil degradation (Oldeman et al., 1991), water shortage and the rising demand for food and fibre to feed growing populations.

3.8 Economy

3.8.1 Economic losses

Key messages

- In Europe, 64 % of all catastrophic events since 1980 are directly attributable to weather and climate extremes: floods, storms and droughts/heatwaves. 79 % of economic losses caused by catastrophic events result from these weather and climate related events.
- Economic losses resulting from weather and climate related events have increased significantly in the last 20 years, from an annual average of less than USD 5 billion to about USD 11 billion. This is due to wealth increase and more frequent events. Four out of the five years with the largest economic losses in this period have occurred since 1997.
- The average number of annual disastrous weather and climate related events in Europe doubled over the 1990s compared with the previous decade, while non-climatic events such as earthquakes remained stable.
- Climate change projections show an increasing likelihood of extreme weather events. Thus, an escalation in damage caused is likely.

Figure 3.26 Weather and climate related disasters in Europe

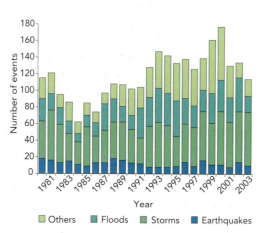

Source: NatCat Service, Munich Re, 2004.

Figure 3.27 Economic and insured losses caused by weather and climate related disasters in Europe

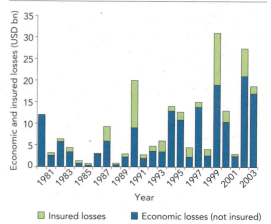

Source: NatCat Service, Munich Re, 2004.

Relevance

Economic losses from extreme weather and climate events, based on information from the financial sector, provide a qualitative indicator of socio-economic impacts of climate change. Changes in the frequency and intensity of storms, floods and droughts due to climate change affect the financial sector, including the insurance sector, via the amount of compensation payments. This indicator can also help to identify which sectors, e.g. agriculture, forestry,

infrastructure, industry or private households, are most affected by damage.

Part of the observed increase in losses is due to socio-economic trends such as population growth, increased wealth, changes in river-basin management and urbanisation in vulnerable areas. Climatic factors, mainly changes in storms, precipitation, flooding and drought, also play an increasing role. Precise attribution of economic losses to each of these two causal factors is complex, and there are differences across regions and according to event type.

Data on the trends in the number of catastrophic events and damage in Europe are subject to various uncertainties, e.g. because the coverage has changed over time. Uncertainties in damage estimation are also due to lack of clarity in the share of non-insured and therefore uncompensated damage.

Past trends

In Europe, 64 % of all catastrophic events since 1980 are directly attributable to weather and climate extremes (floods, storms and droughts/heatwaves) and 25 % are attributable to landslides and avalanches, which are also caused by weather and climate. 79 % of the economic losses and 82 % of all deaths caused by catastrophic events result from these weather and climate related events (Wirtz, 2004). The annual average number of these weather and climate related disastrous events in Europe doubled in the 1990s compared with the previous decade, while non-climatic events, such as earthquakes, remained stable (Figure 3.26).

This European trend is similar to the global trend. Annual global economic losses from catastrophic events increased from USD 4 billion a year in the 1950s to USD 40 billion/yr in the 1990s. The insured portion of these losses rose from a negligible level up to USD 9.2 billion/yr during the same period, with a significantly higher insured fraction in industrialised countries. Global weather and climate related losses between 1985 and 1999 increased by a factor of three in relation to insurance premiums (IPCC, 2001a). This is a clear indication of the increasing vulnerability of the insurance sector to climate change. In Europe, economic losses caused by weather and climate related events have increased during the last 20 years from an annual average of less than USD 5 billion to about USD 11

Source: Jean-Paul Pelissier, Reuters, 2003.

billion. Four of the five years with the largest economic losses in this period have occurred since 1997. The insured portion of the losses generally rose, but the increase varied between the years (Figure 3.27).

A particularly disastrous event was the severe flooding in central Europe in August 2002. Austria, the Czech Republic, Germany, Slovakia and Hungary suffered economic losses of about USD 17.3 billion and insured losses of about USD 4.1 billion (EEA, 2004).

Projections (future trends)

It is likely that future projected climate change will include an increase in the intensity and frequency and a shift in spatial distribution of catastrophic events such as storms, floods and droughts. The possible damage will raise the risk of high economic losses, increasing the vulnerability of the insurance sector. The financial services sector as a whole is expected to be able to cope with the impacts of future climate change, although 'low probability-high impact' events or multiple successive events could severely affect parts of the sector. Trends toward increasing size of companies, greater diversification, greater integration of insurance with

Source: Jean-Paul Pelissier, Reuters, 2003.

other financial services, and improved tools to transfer risks potentially contribute to this robustness. Nonetheless, the property/casualty insurance and reinsurance segments are facing greater risks due to weather and climate related increases in compensation. Small, specialised or un-diversified companies even run the risk of bankruptcy. The banking industry, as a provider of loans, may be vulnerable to climate change under some conditions and in some regions. However, in many cases the banking sector transfers its risk back to the insurers who often purchase debt products (IPCC, 2001).

3.9 Human health

3.9.1 Heatwaves

Key messages

- More than 20 000 excess deaths attributable to heat, particularly among the aged population, occurred in western and southern Europe during the summer of 2003.
- Heatwaves are projected to become more frequent and more intense during the twenty-first century and hence the number of excess deaths due to heat is projected to increase in the future. On the other hand, fewer cold spells will likely reduce the number of excess deaths in winter.

Figure 3.28 Number of reported deaths and minimum and maximum temperature in Paris during the heatwave in summer 2003

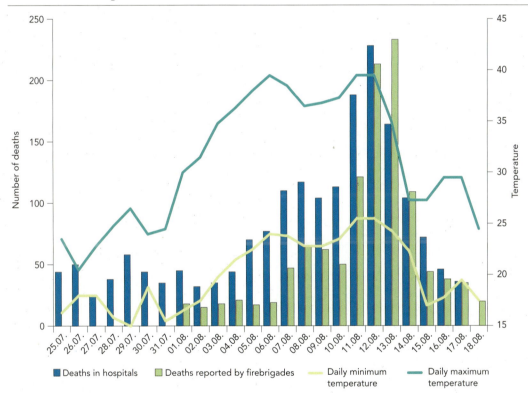

Notes: Reported deaths (bars on left axis), temperature (lines, right axis).
Source: IVS, 2003.

Relevance

Studies (Curriero *et al.*, 2002; Jendritzky *et al.*, 2000; Martens, 1997) have shown that high and low temperatures affect health and well-being. High temperatures can cause clinical conditions such as heat stroke, heat exhaustion, heat syncope, heat eruptions, heat fatigue and heat cramps (WHO, 2004). Many causes of death increase during heatwaves, especially those from cardiovascular and respiratory disease in temperate countries (Faunt *et al.*, 1995; Sartor *et al.*, 1995). The very old and the frail are most susceptible to heat related illness. Heatwaves are particularly dangerous in conjunction with air pollution, e.g. ozone (WHO, 2004). Extremely high temperatures occurring during heatwaves can even lead to a short-term increase in mortality (WHO, 2003b).

Source: Waltraud Grubitzsch, dpa, 2003.

The heatwave in the summer of 2003 resulted in an estimated 20 000 excess deaths, particularly among the aged population in France, Italy, Spain, Portugal and other countries. About 80 % of the 14 000 excess deaths in France between the 1 and 20 August 2003 were of people over 75 years of age (Empereur-Bissonet, 2004). The high night time temperatures (above 25 °C) had particularly negative impacts on human health (Figure 3.28). The heatwave of 2003 showed that many countries are not sufficiently prepared for such events, and that there is a need to take preventive action and to monitor improvements.

It can be difficult for existing social and health systems to cope with heatwaves. Improved emergency management is required to reduce adverse impacts. Data on health impacts of heatwaves are usually acquired by health authorities and rescue services. Uncertainties are relatively low and result from the difficulty of attribution to heatwaves due to other factors that contribute to causes of death. Comparison with average numbers of deaths in recent years in the same season and a statistical analysis of data for the whole year could reduce the uncertainty.

Past trends

Europe has experienced an unprecedented rate of warming in recent decades (see Section 3.2.2). In European cities, overall mortality rises as summer temperatures increase (Katsouyanni *et al.*, 1993; Kunst *et al.*, 1993; Jendritzky *et al.*, 1997). Episodes of extremely high temperatures (heatwaves) also have significant impacts on health. Heatwaves in July 1976 and July–August 1995 were associated with a 15 % increase in mortality in greater London (McMichael and Kovats, 1998; Rooney *et al.*, 1998). A major heatwave in July 1987 in Athens was associated with 2 000 excess deaths (Katsouyanni *et al.*, 1988, 1993).

Projections (future trends)

The current impacts of heatwaves provide a picture of what is likely to happen in the future, since climate models project that the frequency and intensity of heatwaves will further increase (see Section 3.2.4). Even modest changes in average temperature can increase the frequency of heatwaves significantly (IPCC, 2001a, b). The annual excess summertime mortality attributable to climate change is estimated to increase several-fold by 2050 (WHO, 2003). Potentially, in temperate climates, a reduction in (excess) winter deaths caused by the decreasing likelihood of cold spells may outnumber the increase in summer deaths. However, the estimation of the net impact on annual (excess) mortality is currently not possible due to lack of suitable data (WHO, 2003b).

Estimations of future impacts have to take into account whether the population can adapt and acclimatise to changes in climate (Campbell-Lendrum, 2003). Without acclimatisation (physiological, infrastructural and behavioural), the impacts on human health could be severe (WHO, 2003b).

3.9.2 *Flooding*

Key messages
- Between 1975 and 2001, 238 flood events were recorded in Europe. Over this period the annual number of flood events clearly increased.
- The number of people affected by floods rose significantly, with adverse physical and psychological human health consequences.
- Fatal casualties caused per flood event decreased significantly, likely due to improved warning and rescue measures.
- Climate change is likely to increase the frequency of extreme flood events in Europe, in particular the frequency of flash floods, which have the highest risk of fatality.

Figure 3.29 Number of flood events (left); number of deaths per flood event (right)

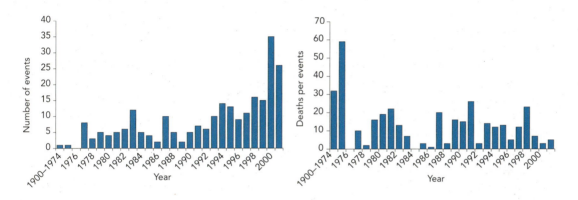

Note: Flood events include flash floods, river floods and storm surges in Europe (1976–2001).
Source: WHO, 2003a.

Relevance

Climate change, including the increasing intensity of heavy rainfall, is likely to make extreme floods more frequent in some areas in Europe. In particular, the number of flash floods is expected to rise which will also increase the risk of casualties. Nevertheless, not all floods and their impacts on human health can be attributed to climate change.

The impacts of the severe floods in August 2002 in Austria, the Czech Republic, Germany, Hungary and the Russian Federation, with more than 100 casualties, together with the increased probability of future flooding, highlight the issue of floods as a risk for human health (WHO, 2003a; EEA, 2004).

The adverse impacts on human health are complex and far-reaching. The number of deaths directly associated with flooding is closely related to the life-threatening characteristic of floods. Deaths are high during sudden flash floods after heavy rainfall, which, in many cases, happen with no prior warning. The death rate is relatively low in the case of river floods or storm surges, which can be forecast several days in advance. Injuries can occur but are more frequent in the aftermath of a flood disaster as residents return to their homes to clean up damage and debris.

Indirect health effects can include those caused by a lack of medical aid due to the damage of major infrastructure. Increases in diseases, such as

Source: Lutz Hennig, 2002.

gastrointestinal diseases, dermatitis and some rare cases of vector borne diseases, have occurred. Most of these illnesses are attributable to reduced sanitation or to overcrowding among displaced people. In addition, poisoning can occur from the rupture of underground pipelines, dislocation of storage tanks, overflow of toxic waste sites or the release of chemicals stored at ground level. Aside from the trauma caused by the flooding itself, many psychological health problems arise from geographical displacement, damage to the home or loss of familiar possessions, and often from lack of insurance.

Data on death and injuries caused by flooding have a relatively high uncertainty. This is due to a lack of information about injuries (mainly only casualties are reported) and ambiguous causes of death by indirect effects, e.g. death during hospital evacuation.

Past trends

Flooding is the most common natural disaster in Europe. The total number of flood events in Europe has increased since 1974, while the number of deaths per flood event has decreased (Figure 3.29), likely due to improved warning and rescue systems. The International Disaster Database recorded 238 floods in Europe between 1975 and 2001. In the past decade, 1 940 people have died during floods and 417 000 have been made homeless.

From January to December 2002, 15 major floods occurred in Europe, e.g. in Austria, the Czech Republic, Germany, Hungary and the Russian Federation. These floods killed approximately 250 people and adversely affected a further one million (WHO, 2003b).

Projections (future trends)

The frequency and intensity of flood events will be closely related to future changes in the pattern of precipitation and river discharge (see Sections 3.2 and 3.5). These changes include widespread increases in rainfall in western and northern Europe, decreases across eastern and southern Europe, and also significant contrasts between winter and summer. It is foreseen that episodes of intense precipitation will grow in frequency, especially in the winter, thus increasing the risk of flooding. In addition, the winter precipitation will fall more often as rain, as a result of higher temperatures. This will lead to immediate runoff and greater risk of flooding (IPCC, 2001b).

The human health impact of future flooding will be strongly determined by the improvement of warning and rescue measures.

3.9.3 Tick-borne diseases

Key messages

- Tick-borne encephalitis cases increased in the Baltic region and central Europe between 1980 and 1995, and have remained high. Ticks can transmit a variety of diseases, such as tick-borne encephalitis (TBE) and Lyme disease (in Europe called Lyme borreliosis).
- It is not clear how many of the 85 000 cases of Lyme borreliosis reported annually in Europe are due to the temperature increase over the past decades.

Map 3.14   Tick prevalence (white dots) in central and northern Sweden

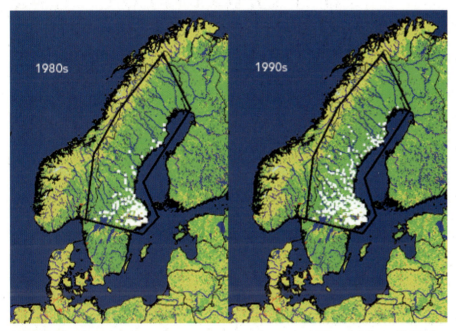

Note: White dots show prevalence, comparing the early 1980s and the mid 1990s in the same region (black line).
Source: Lindgren *et al.*, 2000.

Relevance

Ticks can transmit a variety of diseases, including Lyme borreliosis and tick-borne encephalitis (TBE). These diseases are usually not lethal but can seriously affect health. For quite some time, TBE was regarded as a local problem in some countries. Now it is an infectious disease of increasing importance and reaches epidemic proportions in many parts of Europe and Asia.

The year-round survival, egg hatching and development of ticks from larvae into nymphs and finally adults depend on ecological and climatic conditions.

The altitudinal limit extends up to 2 000 m in southern Europe, 1 300 m in the Italian Alps and 700 m in mountainous areas in central Europe. The number of days per season with temperatures and humidity favourable for tick activity directly affects their abundance. Indirectly, climate influences tick occurrence through effects on their habitat and on the occurrence of host animals. Therefore, changes in the distribution of ticks and tick-borne diseases might indicate changes in climate conditions.

Tick-borne diseases might cause high health care and treatment costs, economic loss through absence from work and a negative perception of tourist destinations. Public health institutes usually monitor the occurrence of ticks. Uncertainties are relatively low and mainly result from incomplete knowledge of the factors which determine tick abundance.

Past trends

An increase in cases of tick-borne diseases per year has been observed since the 1980s in the Baltic countries (Sweden, Finland, Poland, Latvia, Estonia and Lithuania) as well as the central European countries (Switzerland, Germany, Czech Republic and Slovakia).

In Sweden, ticks expanded their northern distribution extensively between 1980 and 1995 (Map 3.14) (Jaenson *et al.*, 1994; Lindgren *et al.*, 2000). During this period, northern areas with newly established tick populations had less severe winters and more summer days. Similar changes have occurred in the Czech Republic with a shift in the upper boundary of the permanent tick population from 700 m to 1 100 m in the period 1982–2002 (Daniel and Kriz, 2002). Over these years the annual mean temperature in the Czech Republic increased by 0.3 °C.

Source: A.R. Walker, University of Edinburgh.

The density of tick populations has increased in endemic areas (permanently populated by ticks) in central Sweden during the same period (Tälleklint and Jeanson, 1998), due to milder winters, earlier springs and prolonged autumns (Lindgren *et al.*, 2000). The chance of winter survival for both ticks and host animals increased as well as the length of the vegetation season. This means easier access to food for host animals and longer periods of activity for ticks.

However, it is not clear yet how many of the 85 000 cases of Lyme borreliosis reported annually in Europe are attributable to temperature increases over the past decades.

Projections (future trends)

The projected further rise in temperature in Europe is likely to increase the geographical extent of ticks and the infestation of areas which are currently tick-free. Due to lack of knowledge of causal relationships between climate change and increases in ticks, tick activity and tick infection rates, projections of future trends have not been possible so far.

Source: www.pixelquelle.de.

4   Adaptation

4.1   Need for adaptation

Even if society substantially reduces its emissions of greenhouse gases over the coming decades, the climate system is projected to continue to change over the coming centuries. Human induced climate change already has, and is expected to continue having, considerable impacts on the environment, human health and various sectors of society. Thus, society has to prepare for and adapt to the consequences of some inevitable climate change, in addition to taking action to mitigate it.

To prevent or limit severe damage to the environment, society and economies, and to ensure sustainable development even under changing climate conditions, adaptation strategies for affected systems are required at European, national, regional and local level.

Adaptation needs the participation of all stakeholders who are involved in any kind of policy, business or service that is or will be affected by climate change. The stakeholder process has to deal with the misconception that adaptation strategies and subsequent actions are always expensive to implement and that non-action is a cheaper alternative.

There are five key reasons why adaptation to climate change is needed and why planning should begin as soon as possible:

(i)   Anticipatory and precautionary adaptation is more effective and less costly than forced, last minute, emergency adaptation or retrofitting.

(ii)   Climate change may be more rapid and more pronounced than current estimates suggest. There is a risk of under-adaptation and the potential for unexpected sudden events.

(iii)   Immediate benefits can be gained from better adaptation to climate variability and extreme climatic events.

(iv)   Immediate benefits can be gained by removing policies and practices that result in ineffective adaptation. An important aspect of adaptive management is to avoid the implementation of decisions that constrain or reduce the effectiveness of future options for adaptation.

(v)   Climate change brings opportunities as well as threats. Future benefits can result from climate change, and these opportunities can be realised or increased by appropriate adaptation and awareness. However, the higher the rate of climate change, the more difficult it will be to realise such benefits or to adapt to its impacts.

4.2   Development of an adaptation strategy

A climate adaptation strategy represents a combination of measures and options chosen to meet a particular risk. Setting up an effective strategy requires several iterative processes (Figure 4.1):

- identifying climate sensitive system components;

- assessing risk;

- identifying potential options for adaptation;

- deciding on and implementing adaptive measures;

Figure 4.1   Decision making framework for climate adaptation strategies

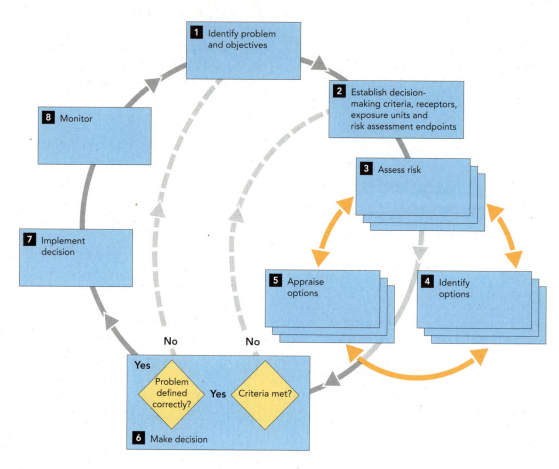

Source: Willows and Connell, 2003.

- ongoing monitoring, assessment and improvement of the implemented measures.

The generic benefits of an adaptation strategy include (Klein and Tol, 1997):

- increasing the robustness of infrastructure designs and long-term investments;

- increasing the flexibility of vulnerable managed systems, e.g. by allowing mid-term adjustments (including changes of activities and location) and reducing economic lifetimes (including increasing depreciation);

- enhancing the adaptability of vulnerable natural systems;

- reversing trends that increase vulnerability to climate;

- improving societal awareness and preparedness.

The success of an adaptation strategy will depend on a combination of factors:

- the flexibility or effectiveness of the measures, including their ability to meet the decision-maker's criteria under a range of climate and non-climate scenarios;

- their potential to produce benefits that outweigh their costs; whether they are consistent with, or complementary to, measures being undertaken by others in related sectors;

- the ease with which they can be implemented.

4.3 Examples of adaptation strategies

There is increasing experience in Europe in implementing national, regional and local strategies for adaptation to climate change, although no European-wide strategy yet exists. Table 4.1 shows a range of examples of adaptation strategies for the climate change impacts dealt with in this report. The applicability of such strategies depends on national, regional and local circumstances.

Table 4.1   Examples of adaptation strategies

Climate change impact	Adaptation strategies (examples)
Increasing temperatures	Modify building design to cope with higher temperatures and enhanced need for summer cooling.
Precipitation extremes	Dams and other flood protection constructions, as well as urban storm sewage systems, need to be assessed for ability to cope with enhanced runoff from more intense storms.
Glacier retreat	Hydro-electric power plants in glacial areas need to adapt to reduced summer flow as glacial extent decreases and enhanced winter flow as temperatures rise.
Snow cover	The skiing industry at lower elevations needs to diversify to take account of decreasing snow cover. Implementation of snow making machines is a short-term strategy.
Arctic sea ice	Need to protect the habitats of indigenous people.
Rise in sea level	Infrastructure changes in vulnerable areas, for example, modification of harbours and ports, enhancement of existing sea defences and managed retreat in areas of low value land.
Marine growing season	Manage changes in fishing and tourism (harmful algae blooms).
Marine species composition	Commercial fishing industry and policy-makers need to take account of shifts in the geographical distribution of species; for example, the northward movement of cod in the southern North Sea.
Species composition	Develop land management strategies that create a porous landscape allowing species to migrate freely.
Plant species distribution in mountain regions	Establish ecological reserves to reduce additional pressure from land-use and tourism activities.
Terrestrial carbon uptake	Strategies to increase carbon sink strength of soils under agricultural land; establish long-rotation tree species; building design to incorporate more wood.
River discharge	Establish flood areas, enforce dikes.
Agriculture	New cropping practices to take account of longer growing season; development of two crops per season agriculture. Establish new varieties; avoid agriculture in risk-zones (flood areas, very dry soils).
Economic losses	Change construction design for buildings and infrastructure; avoid building homes in areas where risks of floods are high.
Human health	Educational campaigns to raise awareness about threats from tick-borne diseases. Campaigns to raise awareness about threats posed by heatwaves.

5 Uncertainties, data availability and future needs

The assessment of climate change and its impacts is still subject to a range of uncertainties and information gaps. For a more comprehensive assessment, more knowledge has to be gathered. This will be done by the IPCC in its next (fourth) assessment due in 2007. This chapter describes the main uncertainties at present, evaluates the data availability for European climate change indicators, and proposes additional indicators that could broaden future European climate change indicator reports.

5.1 Causes of uncertainty

Uncertainties in the assessment of climate change and its impacts result from a cascade of uncertainties regarding all parts of the earth system, from activities leading to emissions, to the responses and adaptation of ecosystems and society. They include:

- Uncertainties about future emissions of greenhouse gases, due to uncertainties in political, demographic, socio-economic and technological development. These uncertainties are managed in the projection of climate change by using different emission scenarios (IPCC, 2000). Together with the uncertainty in climate models, this generates a range of projected future warming (i.e. 1.4–5.8 °C by 2100; IPCC, 2001a).

- Difficulties in attributing an observed change to (human induced) global climate change. There may be an uncertain causal link between an observed change and climate change, and other complex factors may be contributing to the observed change. Time series of the observations may be too short to establish a causal link with high certainty. The frequency and intensity of extreme events are particularly difficult to attribute to

changes in global climate. Biodiversity is affected by climate change, but also by habitat disturbance, fragmentation and destruction due to land-use activities and land-use change. Changes in phenology might be caused by global warming as well as by local changes in climate due to urbanisation (urban sprawl). Thus, it is often hard to determine to what extent natural climate variability, human induced climate change or other anthropogenic factors are responsible for an observed change in a specific indicator.

- Gaps in knowledge about the climate system and hence in climate models. For example, insufficient knowledge about changes in cloud cover and the consequences for the global radiation balance lead to uncertainties in model results about regional precipitation changes. The range of projected temperature and especially precipitation, for example, can be partly attributed to incomplete knowledge of the climate system.

- Insufficient data availability of observed climate change and its impacts. Impacts of climate change show a large variability over time and across Europe, but many datasets are too short or cover too few sample sites to reflect long-term trends and regional variations. As a result, simulation models differ across Europe (see Section 5.2).

In general, much is known about the climate system and the changes in global mean temperature over the past 100 years, whereas less is known about the magnitude of other climate related changes (e.g. the frequency and intensity of extreme events) and the impacts of climate change. Despite these uncertainties, almost all indicators in this report show a clear trend, indicating that the impacts of climate change are already apparent in Europe.

5.2 Data availability

Data availability is one of the main reasons why many climate change indicators are not yet feasible (see Section 5.3). A comprehensive amount of monitoring data across Europe is available for only a few indicators (e.g. temperature, glaciers). Data often only exist for some countries in Europe or even only for regions within countries. For some of the indicators included in this report, the lack of monitoring data made an assessment for the whole of Europe impossible. Instead, indicators had to be based on case studies for a limited number of countries. There is a strong need for improved monitoring and data collection by more intense monitoring programmes. This need is recognised by the EU and many other countries in the second adequacy report of GCOS (Global climate observation system). For many indicators, the required observing techniques will become feasible and cost-effective in the next five to 10 years. For some sectors, particularly the marine sector, considerable research is still required to develop cost-effective technologies. For other indicators, new or additional monitoring programmes have yet to be defined.

To follow is a breakdown of data availability with respect to the indicators presented in this report, and describing how data availability could be improved in the future.

Atmosphere and climate

Climate has been observed over many decades at many locations across Europe. Furthermore, much is known about the climate system and confidence in climate models has increased. However, there is an uncertainty due to lack of knowledge of climate sensitivity, regional climate change, climate variability and extreme events. Changes

in extreme events are important, since these may affect essential aspects of nature and society (e.g. human health), possibly even more than changes in average climate. In addition, projections of precipitation need to be improved, especially in southern and southeastern Europe. A better understanding of future precipitation patterns is needed particularly for this part of Europe because of its sensitivity to water shortage, heatwaves and other extreme events. Due to uncertainties in climate sensitivity, it is unclear which greenhouse gas stabilisation levels would be required to limit temperature rise. For instance, assessments about the GHG concentration level, which would be necessary to limit the temperature rise to the EU target of not more than 2 °C, vary between 550 and 650 parts per million CO_2 equivalents.

Glaciers, snow and ice

Data availability differs among the three selected indicators. For Arctic sea ice, only a limited amount of long-term monitoring data is available. Other data exist, but are not accessible, e.g. due to a military background (sea ice thickness has mainly been observed by military submarines). In the future, Arctic monitoring data are expected to become more easily available, since civil research programmes have recently been carried out. These measurement programmes will soon be supported by satellite programmes like 'Cryosat'.

Satellites have observed snow cover for about 40 years, in addition to terrestrial monitoring programmes in individual countries (e.g. Switzerland, Norway). Detailed investigations have recently begun in Switzerland, driven by the importance of snow for tourism. However, there is a lack of data in other regions on the changing extent in area and altitude of snow packs.

Data availability for glacier mass and length is comparatively good. Large monitoring datasets have been recorded in the last few decades by intensive observation programmes at various national and international institutions. These data are archived centrally at the World Glacier Monitoring Service in Zurich.

Further research is needed to compile indicators describing the climate change effect on permafrost regions. Although such regions are rare in Europe, permafrost is an important factor even for the European climate. Permafrost regions contain large amounts of carbon. Thawing of the permafrost could therefore lead to significant emissions of greenhouse gases and thus accelerate climate change. Furthermore, the thawing could have local effects in the permafrost areas of mountain regions and northern Europe, e.g. threatening infrastructure, which is built on frozen ground. Such effects have already been observed in parts of Siberia and Alaska.

Marine systems

High quality sea-level data have been recorded in Europe for more than 100 years. Longer historical records are available for a few tide gauges. Uncertainties in ocean models are still substantial, however, because of an incomplete understanding of the complex ocean system. For this reason, there is a strong need to continue existing and to start new measurement programmes and modelling activities.

Ships have measured sea surface temperatures of the European seas for many decades. Data exist in many national institutes and are aggregated at the International Council for the Exploration of the Sea (ICES) in Copenhagen.

The 'Continuous Plankton Recorder' of the Sir Alister Hardy Foundation for Ocean Science (SAHFOS) in Plymouth (UK) is one of the few research programmes observing the changes in the seasonal and spatial distribution of zoo- and phytoplankton. Despite the high quality of these data, the spatial and temporal extent is limited. Future

activities should focus on an extension of the sampling network and on the development of models, which can estimate the plankton distribution in the oceans.

Terrestrial ecosystems and biodiversity

Phenological data have been collected for many years in different places in Europe. Research is now integrated through the European Phenological Network. However, this activity needs to be intensified to achieve a more reliable attribution of phenological data to climate change, especially at regional level. More regions should be included in order to achieve a Europe-wide picture.

For other indicators related to terrestrial ecosystems, the availability and usefulness of data are often limited. For species composition, for instance, some datasets exist, but these are based on relatively short and small-scale research activities (commonly in plant ecology) and derived from research projects in a limited number of countries in Europe (or even only regions within countries). The attribution of observed changes in these indicators to climate change is improving, although various other causes of changes also exist. More research on the impacts of climate change on biodiversity and ecosystems is needed, i.e. more monitoring programmes and more models on a regional scale.

Projections are subject to additional uncertainties. Firstly, the models in use have an uneven coverage of species distribution across Europe. Secondly, plant species, especially in mountain regions, often cover only small areas, whereas the model input (e.g. climate) is averaged over larger scales. This leads to an underestimation of the distribution of mountain plant species. Thirdly, it remains unclear how terrestrial ecosystems will adapt to rapid climate change. Tolerance ranges of plant species might change in the future. An increase in the annual temperature of

1–2 °C, for example, might be within the range of tolerance of many mountain species and thus only result in a limited composition change. However, the projected warming within the next 100 years will most likely be outside the tolerance range of most plant species, and will thus lead to considerable species losses (Theurillat *et al.*, 1998). The issue of adaptation to climate change has only recently been taken up in ecological research.

Water

A surplus or shortage of water can lead to floods or droughts, disrupting the environment and society. Because of its importance, annual river runoff and other water related variables have been observed for many decades. Several hundred runoff measuring points exist globally, about 70 in Europe. The uncertainty of these datasets is relatively low compared with other components of the hydrological cycle (Kaspar, 2004) because annual river runoff can be measured easily and the process is well understood. Less obvious and therefore more uncertain is the detection of climate change impacts in observed river discharge data. Various, non-climate related factors such as river management play a role. Selecting certain rivers and focusing on annual average changes may reduce this uncertainty. Furthermore, the attribution to observed changes in climate is complex because annual river discharge depends on changes in both precipitation and temperature. Both factors are projected to change differently in space and time. Finally, the uncertainty in projecting future trends is significant, because of incomplete knowledge of factors such as precipitation and regional changes in Europe. Considerable differences between the climate models still exist in projections of precipitation, especially in southern Europe.

Agriculture

Much is known about how climate change affects crop growth in general. This knowledge is based on field experiments, for instance free-air CO_2 enrichment experiments (FACE), and on model based simulations that can simulate the physiological reaction of plants to changing conditions. This knowledge has been gained on a local scale but also integrated on a European scale (due to the JRC-MARS project, http://projects-2001.jrc.cec.eu.int). All studies show that climate is an important driving factor of crop yield. However, the contribution of climate change to observed trends in crop yield is subject to uncertainties due to complex interactions between technical progress, policy measures and crop yield in the EU. This is also valid for the projection of future trends. Furthermore, uncertainties in the projection of future precipitation complicate the estimates of future yield gains or losses. This is particularly true for warm and dry areas (e.g. southern and southeastern Europe), where water will be a critical factor for agriculture in the future. In these areas, model results diverge to a great extent, depending on the scenarios in use and the model itself. While some models project large increases in yield for southern Europe due to the positive effect of increasing CO_2 concentration, others project losses as a consequence of insufficient water supply due to rising temperatures and decreasing rainfall. For central and northern Europe, where water supply is less critical, projections are relatively robust.

Economy

There is only limited information about the impacts of climate change on the economic sector.

Data on the trends in the number of catastrophic events as well as on the human casualties and economic losses in Europe are collected by the International Disaster Database (EM-DAT) in Belgium and by insurance and re-insurance companies (e.g. Munich Re). They give an overview of trends in damage to the different economic sectors, including industry, energy, infrastructure and private households. Other impacts are mostly related to micro-economic processes, e.g. tourism or energy consumption per capita due to heatwaves or extremely cold periods. In most cases, economic processes are also important and, therefore, it is difficult to separate climatic from non-climatic effects.

Further efforts and more detailed datasets are necessary to evaluate impacts of climate change on economic processes. Energy consumption and tourism seem to be the most promising issues for further investigation.

Human health

Developing indicators for human health in relation to climate change is a relatively new research field. In 2000, the WHO/ECEH (European Centre for Environment and Health, Rome) started a programme called 'Climate change and human health'. This programme focuses on changes of allergic disorders, water and food-borne diseases, vector-borne diseases and human health impacts due to catastrophic events such as floods or heatwaves. The indicators on effects of heatwaves and floods as well as on tick-borne diseases have been included in this report using data and information from the WHO. There is still a lack of understanding of many processes and interactions between climate and human health, and a lack of data (e.g. on the number of climate related excess deaths).

A continuous cooperation between environment agencies, including the EEA, and health organisations, including the WHO/ECEH, is essential to improve the understanding of the causal relationship between climate change and health.

5.3   Need for additional indicators

The 22 indicators presented in this report cover a broad range of climate state and impact categories. Although the objective of the report was to show impacts from a broad perspective, the indicators do not provide a complete picture of all climate change impacts in Europe that are currently occurring. Future work should update and broaden the approach of this report.

Important impacts like floods and droughts, which are often related to extreme events, could only be partly included, mainly due to limitations in data availability. Impacts of climate change on forestry and economics are mixed with other human activities. More detailed analyses are necessary to separate climatic from non-climatic effects.

Indicators on the impact of climate change on socio-economic processes are particularly relevant for the general public, stakeholders and decision-makers, who have to understand the threats of climate change in order to define useful adaptation strategies. Defining such strategies is a crucial issue for the next IPCC assessment report (to be published in 2007) and in the climate change policy debate.

A number of further indicators, which cover some of these additional categories, were already identified as potentially useful and described in an EEA technical report (see EEA, 2002). Indicators, which could be feasible for reporting in the next five years, are listed in Table 5.1. As more information becomes available in the future, some of these indicators could be reconsidered to achieve a more comprehensive picture of climate change impacts in Europe.

Many of the indicators included in this report and the proposed new indicators are also included in a list of essential climate variables that have been identified by the global climate observing system (GCOS) and accepted by all parties, including the EU, to the UN climate convention (GCOS, 2003). The GCOS list includes a more extensive range of indicators, but most of these require long-term measurement programmes (GCOS developed a five

Table 5.1   List of possible additional climate change state and impact indicators and possible (near-term) sources of data. (The list is a result of two expert meetings (EEA, 2002); the valuation is according to the authors' experience)

Category	Proposed indicators not included in this report	Attribution to climate change (*)	Data availability (**)
Atmosphere and climate	Climate indices	++	+
	Upper stratosphere temperature	++	+
	Storms, storm surges and other extreme events	+	o
Glaciers, snow and ice	Lake and river ice	++	+
	Changes in permafrost	+	+
Agriculture and forestry	Forest suitability	++	+
	Forest growth	+	++
	Shifts in tree lines	+	o
	Pests and diseases	+	o
	Crop suitability	++	+
Ecosystems and biodiversity	Changes in species behavioural patterns	+	o
Hydrology and water resources	Frequency of low/high river flows	++	+
	Lake temperatures	+	+
	Fresh water availability	+	+
Marine environment and coastal zones	Characteristics of storm surges	+	+
	Thermohaline circulation	+	+
	Coastal erosion, retreat	+	+
Economy and infrastructure	Energy consumption for heating in winter	++	+
	(Winter) tourism	++	+
Human health	Catastrophic weather related injuries	+	+
	Seasonal peak of food-/water-borne diseases	o	o
	Seasonal change of allergic pollen	+	o

(*) Can the indicator be attributed to climate change (active or passive)?
++ very well attributable
+   attribution possible, indicator is more complex, also driven by other forces
o   attribution is unclear or unknown
(**) Could the indicator be described sufficiently by existing data?
++ sufficient data exist
+   data exist, but possibly not sufficient
o   data very limited, sufficient description unlikely

to 10-year implementation plan to fill some of these gaps). GCOS believes that there is strong need for data collection in assessing climate change and its impacts. The improved data collection comprises:

- more intense observation programmes and networks, both ground measurements and satellite observations, supplemented by new simulation models. Climate observation in the terrestrial domain remains the least well-developed component of GCOS, although increasing significance is being placed on terrestrial data;

- a clear data definition and standardisation, including provision of meta/background data;

- use of historical data (e.g. in digital format) to establish long-term datasets;

- ensuring free access to and unrestricted exchange of data. Arctic sea ice data are, for example, available in large quantities but not accessible because of their military background;

- collecting all data at central data centres;

- extending capacity building and training particularly in the least-developed countries and countries in transition.

Table 5.2   Essential climate variables that are currently available for global implementation and have a high impact on UNFCCC requirements

Domain	Essential climate variables	
Atmospheric (over land, sea and ice)	Surface:	Air temperature, precipitation, air pressure, surface radiation budget, wind speed and direction, water vapour.
	Upper-air:	Earth radiation budget (including solar irradiance), upper-air emperature (including MSU — microwave sounding unit — radiances), wind speed and direction, water vapour, cloud properties.
	Composition:	Carbon dioxide, methane, ozone, other long-lived greenhouse gases, aerosol properties.
Oceanic	Surface:	Sea-surface temperature, sea-surface salinity, sea level, sea state, sea ice, current, ocean colour (for biological activity), carbon dioxide partial pressure.
	Sub-surface:	Temperature, salinity, current, nutrients, carbon, ocean tracers, phytoplankton.
Terrestrial		River discharge, water use, ground water, lake levels, snow cover, glaciers and ice caps, permafrost and seasonally frozen ground, albedo, land cover (including vegetation type), fraction of absorbed photosynthetically active radiation (FAPAR), leaf area index (LAI), biomass, fire disturbance.

Source: GCOS, 2003.

References

1. Introduction

Hulme, M., Jenkins, G.J., Lu, X., Turnpenny, J.R., Mitchell, T.D., Jones, R.G., Lowe, J., Murphy, J.M., Hassell, D., Boorman, P., McDonald, R. and Hill, S. (2002): Climate change scenarios for the United Kingdom: The UKCIP02 scientific report, Tyndall Centre for Climate Change Research, Norwich, UK.

IPCC (2001a): Climate change 2001: The scientific basis, Cambridge University Press, Cambridge, UK.

IPCC (2001b): Climate Change 2001: Impacts, adaptation and vulnerability, IPCC WGII report, Cambridge University Press, Cambridge, UK.

Sweeney, J., Donnelly, A., McElwain, L. and Jones, M. (2002): Climate change indicators for Ireland. EPA — Irish Environmental Protection Agency, Wexford, Ireland.

2. Background

CBD — Convention on Biological Diversity (2003): Linkages between biological diversity and climate change and advice on the integration of biodiversity considerations into the UNFCCC and its Kyoto Protocol. CBD ad hoc technical expert group on biodiversity and climate change. Draft.

CRU (2003): Global average temperature change 1856–2003. http//www.cru.uea. ac.uk/cru/data/temperature/

DTI (2003a): Our energy future — creating a low carbon economy, Energy White Paper, UK.

DTI (2003b): Options for a low carbon economy, DTI economics paper No 4, UK.

EEA (2003a): Europe's environment: the third assessment, Environmental assessment report No 10, Copenhagen, Denmark.

European Parliament and Council (2002): Decision No. 1600/2002/EC, laying down the sixth community environment action programme, 22 July 2002.

IPCC (2001a): Climate change 2001: The scientific basis, Cambridge University Press, Cambridge, UK.

IPCC (2001b): Climate change 2001: Impacts, adaptation and vulnerability, IPCC WGII report, Cambridge University Press, Cambridge, UK.

Mann, M.E., Bradley, R.S. and Hughes, M.K. (1999): 'Northern hemisphere temperatures during the past millennium: inferences, uncertainties, and limitations', Geophysical Research Letters, 26.

Petit, J.R., Jouzel, J., Raynaud, D., Barkov, N.I., Barnola, J.-M., Basile, I., Benders, M., Chappellaz, J., Davis, M., Delayque, G., Delmotte, M., Kotlyakov, V.M., Legrand, M., Lipenkov, V.Y., Lorius, C., Pépin, L., Ritz, C., Saltzman, E. and Stievenard, M. (1999): 'Climate and atmospheric history of the past 420,000 years from the Vostok ice core, Antarctica', Nature, 399, pp. 429–436.

UNFCCC (1993): The United Framework Convention on Climate Change.

UNFCCC (2003): National communications from Parties included in Annex I to the convention. Compilation and synthesis report. FCCC/SBI/2003/7 and Add.1–4. UNFCCC secretariat. Bonn, Germany.

UKCIP (2002): Climate change scenarios for the United Kingdom, Oxford, UK.

WBGU — German Advisory Council on Global Change (2003a): Climate protection strategies for the 21st Century: Kyoto and beyond, Special Report 2003, WBGU, Berlin, Germany.

WBGU — German Advisory Council on Global Change (2003b): World in transition: Towards sustainable energy systems, Berlin, Germany.

WHO — World Health Organisation (2003): Climate change and human health, WHO.

3. Climate change impacts in Europe

Introduction

Cannell, M.G.R., Palutikof, J.P. and Sparks, T.H. (eds) (1999): Indicators of climate change in the UK, Department for Environment, Transport and the Regions, London, UK.

Cannell (2003): Review of UK climate change indicators, study for DEFRA (Department for Environment Food and Rural Affairs), June 2003.

EEA (2002a): Environmental signals 2002, Environmental assessment report No 9, European Environment Agency, Copenhagen, Denmark.

EEA (2002b): Proposed set of climate change state and impact indicators in Europe. Technical report (ETC/ACC) European Environment Agency, Copenhagen, Denmark.

EEA (2003a): Greenhouse gas emission trends and projections in Europe, Environmental issue report No 36, European Environment Agency, Copenhagen, Denmark.

EEA (2003b): Europe's environment: the third assessment, Environmental assessment report No 10, European Environment Agency, Copenhagen, Denmark.

EEA (2003c): EEA core set of indicators, European Environment Agency, Copenhagen, Denmark.

EEA (2004): Mapping the impacts of recent natural disasters and technological accidents in Europe, Environmental issue report No 35, European Environment Agency, Copenhagen, Denmark.

European Commission (2003a): Second ECCP progress report — Can we meet our Kyoto targets?, Brussels.

Freibauer *et al.* (2002): CarboEurope, a cluster of projects to understand and quantify the carbon balance of Europe, CarboEurope European Office, Jena, Germany.

Hadley Centre (2003): Climate change observations and predictions, December 2003, UK.

Hulme, M., Jenkins, G.J., Lu, X., Turnpenny, J.R., Mitchell, T.D., Jones, R.G., Lowe, J., Murphy, J.M., Hassell, D., Boorman, P., McDonald, R. and Hill, S. (2002): Climate change scenarios for the United Kingdom: The UKCIP02 scientific report, Tyndall Centre for Climate Change Research, Norwich, UK.

IEA (2002): Dealing with climate change: Policies and measures in IEA member countries, OECD/IEA.

IPCC (2001a): Climate change 2001: The scientific basis, Cambridge University Press, Cambridge, UK.

IPCC (2001b): Climate change 2001: Impacts, adaptation and vulnerability, IPCC WGII report, Cambridge University Press, Cambridge, UK.

Parry, M.L. (ed) (2000): Assessment of potential effects and adaptation for climate change in Europe: The Europe Acacia Project, Jackson Environmental Institute, University of East Anglia, Norwich, UK.

UNFCCC (2003): National communications from Parties included in Annex I to the convention. Compilation and synthesis report. FCCC/SBI/2003/7 and Add.1–4. UNFCCC secretariat. Bonn, Germany.

UNFCCC (2004): Compendium on methods and tools to evaluate impacts of, vulnerability and adaptation to climate change. UNFCCC, final draft. http://unfccc.int/program/mis/meth/compendium.pdf

WBGU — German Advisory Council on Global Change (2003): Climate protection strategies for the 21st Century: Kyoto and beyond, Special Report 2003, WBGU, Berlin, Germany.

WHO (2003): Climate change and human health? Risks and responses, WHO/WMO/UNEP, Geneva, Switzerland.

Atmosphere and climate

CRU (2003): Global average temperature change 1856–2003. http//www.cru.uea.ac.uk/cru/data/temperature/

European Parliament and Council (2002): Decision No. 1600/2002/EC, laying down the sixth community environment action programme, 22 July 2002.

Gillett, N. P., Graf, H. F. and Osborn, T. J. (2003): Climate change and the NAO, American Geophysical Union, Washington DC, USA.

Hurrell, J.W. (1996): 'Influence of variations in extratropical wintertime teleconnections on northern hemisphere temperature', Geophysical Research Letters, 23 (6), pp. 665–668.

IPCC (2001a): Climate change 2001: The scientific basis, Cambridge University Press, Cambridge, UK.

IPCC (2001b): Climate change 2001: Impacts, adaptation and vulnerability, IPCC WGII report, Cambridge University Press, Cambridge, UK.

Jones, P.D., New, M., Parker, D.E., Martin, S. and Rigor, I.G. (1999): 'Surface air temperature and its changes over the past 150 years', Review of Geophysics, 37, pp. 173–100.

Jones, P.D. and Moberg, D. (2003): Hemispheric and large-scale surface air temperature variations: An extensive revision and an update to 2001, Journal of Climate, 16, pp. 206–223.

Klein Tank, A., Wijngaard, J. and van Engelen, A. (2002): Climate in Europe. Assessment of observed daily temperature and precipitation extremes. European Climate Assessment, KNMI, the Bilt, the Netherlands. See also http://www.knmi.nl/samenw/eca/

Leemans, R. and Hootsmans, R. (1998): Ecosystem vulnerability and climate protection goals, RIVM report 481508004, Bilthoven, the Netherlands.

Luterbacher, J., Dietrich, D., Xoplaki, E., Grosjean, M. and Wanner, H. (2004): 'European seasonal and annual temperature variability. Trends and extremes since 1500', Science 303, pp. 1499–1503.

NOAA — National oceanic and atmospheric administration climate monitoring and diagnostics laboratory (2003): (CMDL)/Halocarbons and other atmospheric trace species (HATS). Data are made available through http://www.cmdl.noaa.gov/hats/

Parry, M.L. (ed) (2000): Assessment of potential effects and adaptation for climate change in Europe: The Europe Acacia Project, Jackson Environmental Institute, University of East Anglia, Norwich, UK.

Rijsberman, F. and Swart, R.J. (eds) (1990): Targets and indicators of climate change, Stockholm Environment Institute, 166.

Romero, R., Ramis, C. and Guijarro, J.A. (1999): 'Daily rainfall patterns in the Spanish Mediterranean area: an objective classification', International Journal of Climatology, 19, pp. 95–112.

Schär, C., Vidale, P.L., Lüthi, D., Frei, C., Häberli, C., Liniger, M.A. and Appenzeller, C. (2004): The role of increasing temperature variability in European summer heatwaves, Nature 427, pp. 1–4.

WBGU — German Advisory Council on Global Change (2003b): World in transition: Towards sustainable energy systems, Berlin.

WMO, 2003. Global temperature in 2003, WHO press release, December 2003.

Glaciers, snow and ice

Armstrong, R.L. and Brodzik, M.J. (2001): 'Recent northern hemisphere snow extent: a comparison of data derived from visible and microwave sensors', Geophysical Research Letters, 28(19), pp. 3673–76.

Beniston, M. *et al.* (1995). 'Simulation of climate trends over the Alpine region', in: Development of a physically-based modelling system for application to regional studies of current and future climate, Final scientific report No. 4031–33250, Swiss National Science Foundation, Bern, Switzerland.

Bjorgo, E., Johannessen, O.M. and Miles, M.W. (1997): Analysis of merged SMMR SSMI time series of Arctic and Antarctic sea ice parameters, 1978–1995, Geophysical Research Letters, 24, pp. 413–416.

Bürki, R. *et al.* (2003): Climate change and winter sports: Environmental and economic threats, 5th IOC/UNEP World Conference on Sport and Environment, Turin, Italy, December 2003.

Cavalieri, D.J., Gloersen, P., Parkinson, C.L., Comiso, J.C. and Zwally, H.J. (1997): 'Observed hemispheric asymmetry in global sea ice changes', Science, 278, pp. 1104–1106.

Dye, D. (1997): Satellite analysis of inter-annual variability and trends in the northern hemisphere annual snow-free period, Department of Geography, Boston University, USA.

Dyurgerov, M. (2003): 'Mountain and subpolar glaciers show an increase in sensitivity to climate warming and intensification of the water cycle', Journal of Hydrology 282, pp. 164–176.

Frauenfelder, R., (2003): Personal Communication, WGMS.

Haas, C. (2004): 'Late-summer sea ice thickness variability in the Arctic Transpolar Drift 1991–2001 derived from ground-based electromagnetic sounding', Geophysical Research Letters, in press.

Haeberli, W. (2003): Spuren des Hitzesommers 2003 im Eis der Alpen, Submission to the parliament of Switzerland, 30 September 2003.

Haeberli, W. and Beniston, M. (1998): Climate change and its impacts on glaciers and permafrost in the Alps, Ambio Vol. 27 No 4, pp. 258–265.

IPCC (2001a): Climate change 2001: The scientific basis, Cambridge University Press, Cambridge, UK.

Johannessen, O.M., Miles, M.W. and Bjørgo, E. (1995): The Arctic's shrinking sea ice, Nature, 376, pp. 126–127.

Johannessen, O.M., Bengtsson, L.B. *et al.*, (2002): Arctic climate change — Observed and modelled temperature and sea ice variability, NERSC, Technical Report No 218, Bergen, Norway.

Laternser, M. and Schneebeli, M. (2001): 'Climate trends from homogeneous long-term snow data of the Swiss Alps (1931–1999)'. Submitted to International Journal of Climatology.

Loewe, P. and Wilhelms, S. (2002): Personal information, Federal Maritime and Hydrographic Agency, Hamburg, Germany.

Maisch, M. and Haeberli, W. (2003): ‚Die rezente Erwärmung der Atmosphäre-Folgen für die Schweizer Gletscher', Geographische Rundschau 55 Heft 2.

Moritz, R.M., Bitz, C.M. and Steig, E.J. (2002): 'Dynamics of recent climate change in the Arctic', Science, Vol. 297, pp. 1497–1502.

Overland, J.E. *et al.* (2004): Integrated analysis of physical and biological pan-Arctic change, Climate Change 63, pp. 323–336.

Robinson, D.A. (1997): 'Hemispheric snow cover and surface albedo for model validation', Ann. Glaciol., 25, pp. 241–245.

Rothrock, D.A., Yu, Y. and Maykut, G.A. (1999): 'Thinning of the Arctic sea-ice cover', Geophysical Research Letters, 26, pp. 3469–3472.

UKMO (2004): Rayner, Nick; UK Met Office; personal communication.

Wallinga, J. and van de Wal, R.S.W. (1998): 'Sensitivity of Rhonegletscher, Switzerland, to climate change: experiments with a one-dimensional flow-line model', Journal of Glaciology, 44, pp. 383–393.

WMO (2002): Statement on the status of the global climate in 2001, WMO — No 940, 2002.

Marine systems

Beaugrand, G., Ibañez, F., Lindley, J.A. and Reid, P.C. (2002): 'Diversity of calanoid copepods in the North Atlantic and adjacent seas: species associations and biogeography', Marine Ecology Progress Series, 232, pp. 179–195.

Beaugrand, G., Reid, P.C., Ibanez, F., Lindley, J.A. and Edwards, M. (2002): 'Reorganisation of North Atlantic marine copepod biodiversity and climate', Science, 296, pp. 1692–1694.

Cane, M.A. *et al.* (1997): 'Twentieth-century sea surface temperature trends', Science, Vol.275, pp. 957–960.

Dooley, H. (2003): ICES Copenhagen, Scientific contribution and personal information.

Edwards, M. *et al.* (2003): Fact sheet on phytoplankton, submitted to ETC/ACC.

Edwards, M., Reid, P.C. and Planque, B. (2001): 'Long-term and regional variability of phytoplankton biomass in the Northeast Atlantic (1960-1995)', ICES Journal of Marine Science, 58, pp. 39–49.

Edwards, M., Beaugrand, G., Reid, P.C., Rowden, A. and Jones, M. (2002): 'Ocean climate anomalies and the ecology of the North Sea', Marine Ecology Progress Series, 239, pp. 1–10.

Furewik, T. (2000): 'On anomalous sea surface temperatures in the Nordic Seas', Journal of Climate, Vol. 13, pp. 1044–1053.

IPCC (2001a): Climate change 2001: The scientific basis, Cambridge University Press, Cambridge, UK.

IPCC (2001b): Climate change 2001: Impacts, adaptation and vulnerability, IPCC WGII report, Cambridge University Press, Cambridge, UK.

Levitus, S. *et al.* (2000): 'Warming of the world ocean', Science, Vol. 287. pp. 2225–2229.

Liebsch, G., Novotny, K. and Dietrich, R. (2002): Untersuchung von Pegelreihen zur Bestimmung der Änderung des mittleren Meeresspiegels an den europäischen Küsten, Technische Universität Dresden (TUD), Germany.

Melsom, A. (2001): Are North Atlantic SST anomalies significant for the Nordic Seas SSTs?, Norwegian Meteorological Institute.

Mizoguchi, K.I. *et al.* (1999): Multi- and quasi- decadal variations of sea surface temperature in the North Atlantic. Journal of Climate, Vol. 12, pp. 3133–3143.

Nerem, R.S. and Mitchum, G.T. (2001): 'Sea level change', in: Satellite altimetry and Earth sciences, Academic Press, San Diego, USA., pp. 329–349.

Parmesan, C. and Yohe, G. (2003): 'A globally coherent fingerprint of climate change impacts across natural systems', Nature, 421, pp. 37–42.

Peltier, W. (1998): 'Postglacial variations in the level of the sea: Implications for climate dynamics and solid-earth', Review of Geophysics 36(4), pp. 603–689.

Reid, P.C., Edwards, M., Hunt, H.G. and Warner, A.J. (1998): Phytoplankton change in the North Atlantic. Nature 391, p. 546.

Reid, P.C. and Edwards, M. (2001): 'Long-term changes in the pelagos, benthos and fisheries of the North Sea', in: Kröncke, I., Türkay, M. and Sündermann, J. (eds): 'Burning issues of North Sea ecology, Proceedings of the 14th International Senckenberg Conference 'North Sea 2000', Senckenbergiana Maritima, 31, pp. 107–115.

Rixen, M. *et al.* (2004) : 'Western and Eastern Mediterranean Sea temperature and salinity ; variability and trends since 1950', Nature, submitted.

SAHFOS — Sir Alister Hardy Foundation for Ocean Science (2002): Annual Report 2002.

Westernhagen, v. H and Schnack, D. (2001): 'The effect of climate on fish populations', Climate of the twenty-first century, pp. 283–289.

Terrestrial ecosystems and biodiversity

Bakkenes, M., Alkemade, J.R.M., Ihle, F., Leemans, R. and Latour, J.B., (2002): 'Assessing effects of forecasted climate change on the diversity and distribution of European higher plants for 2050', Global Change Biology, 8, pp. 390–407.

Bakkenes M., Eickhout, B. and Alkemade, R. (2004): 'Impacts of climate change on biodiversity in Europe; implication of CO_2 stabilisation scenarios', Global Change Biology, in preparation.

Bousquet, P., Peylin, P., Ciais, P., Le Quere, C., Friedlingstein, P. and Tans, P.P. (2000): Interannual changes in regional CO_2 fluxes, Science 290, pp. 1342–1346.

CBD — Convention on Biological Diversity (2003): Linkages between biological diversity and climate change and advice on the integration of biodiversity considerations into the UNFCCC and its Kyoto Protocol, CBD ad hoc technical expert group on biodiversity and climate change. Draft 17.

Chuine I. and Beaubien, E.G. (2001): 'Phenology is a major determinant of tree species range'. Ecological Letters 4, pp. 500–510.

EEA (2004): Greenhouse gas emission trends and projections in Europe 2004, EEA report, European Environment Agency, Copenhagen, Denmark. (forthcoming)

Frederiksen, M. (2002): 'The use of data from bird ringing schemes as indicators of environmental change: a feasibility study'. ETC/NPB, MNHN/CRBPO.

Gottfried, M., Pauli, H. Reiter, K. and Grabherr, G. (1999): 'A fine-scaled predictive model for changes in species distribution patterns of high mountain plants induced by climate warming'. Diversity Distribution 5: pp. 241–251.

Grabherr, G., Gottfried, M. and Pauli, H. (1994): Climate effects on mountain plants. Nature, pp. 369–448.

Grabherr, G., Gottfried, M. and Pauli, H. (2002): 'Ökologische Effekte an den Grenzen des Lebens', Spektrum der Wissenschaft, 1, pp. 84–89.

Hare, W. (2003): Assessment of knowledge on impacts of climate change, Contribution to the specification of Art. 2 of the UNFCCC. Background report to the WBGU special report 94.

Holten, J.I. and Carey, P.D. (1992): Responses of climate change on natural terrestrial ecosystems in Norway, NINA Forskningsrapport 29, Norwegian Institute for Nature Research, Trondheim, Norway, 59 pages.

Hughes, L. (2000): 'Biological consequences of global warming: is the signal already apparent?', Trends in Ecology and Evolution 15(2), pp. 56–61.

IPCC (2001a): Climate change 2001: The scientific basis, Cambridge University Press, Cambridge, UK.

IPCC (2001b): Climate change 2001: Impacts, adaptation and vulnerability, IPCC WGII report, Cambridge University Press, Cambridge, UK.

IPCC (2002): Climate change and biodiversity; Technical paper 5, IPCC, April 2002.

Janssens I.A., Freibauer, A., Ciais, P., Smith, P., Nabuurs, G.J., Folberth, G. Schlamadinger, B., Hutjes, R.W.A., Ceulemans, R., Schulze, E.-D., Valentini, R. and Dolman, A.J. (2003): 'Europe's terrestrial biosphere absorbs 7–12 % of European anthropogenic CO_2 emissions', Science Express 300 (5623), 22 May 2003. 0.1126/science.1083592.

Klanderud, K. and Birks, H. J. B. (2003): 'Recent increases in species richness and shifts in altitudinal distributions of Norwegian mountain plants'. The Holocene, 13, pp. 1–6.

Körner, C. (1999): Alpine plant life, Springer Press. 338 pages.

Kozlov M.V. and Berlina, N.G. (2002): 'Decline in length of the summer season on the Kola peninsula Russia', in Climate change 54, pp. 387–398.

Kullman, L. (2003): 'Recent reversal of neoglacial climate cooling trend in the Swedish Scandes as evidenced by birch tree-limit rise', Global and Planetary Change, 36, pp. 77–88.

Menzel, A. (2002): 'Phenology: Its importance to the global change community. An editorial comment', Climatic Change 54, pp. 379–385.

Menzel A. and Fabian, P. (1999): 'Growing season extended in Europe', Nature 397, p. 659.

Mitchell, T.D., Carter, T.R. Jones, P.D. Hulme, M. and New, M. (2004): 'A comprehensive set of high-resolution grids of monthly climate for Europe and the globe: the observed record (1901–2000) and 16 scenarios (2001–2100)', Journal of Climate: submitted.

Meshinev, T., Apostolova, I. and Koleva E. (2000): 'Influence of warming on timberline rising: a case study on Pinus peuce in Bulgaria', Phytocoenologia 30, pp. 431–438.

Molau, U. and Alatalo, J.M. (1998): 'Responses of sub-alpine plant communities to simulated environmental change', Ambio 27, pp. 322–329.

Motta, R. and Masarin, F. (1998): 'Strutture e dinamiche forestali di popolamenti misti di pino cembro (*Pinus cembra L.*) e larice (*Larix decidua Miller*) in alta valle Varaita (Cuneo, Piemonte)'. Archivio Geobotanico 2, pp. 123–132.

Nabuurs, G.J., Pussinen, A., Karjalainen, T., Erhard, M., and Kramer, K. (2002): 'Stemwood volume increment changes in European forests due to climate change — a simulation study with the EFISCEN model', Global Change Biology 8, pp. 304–316.

New, M., Hulme, M. and Jones, P. (2000): Representing twentieth-century space-time variability. Part II: Development of 1901–96 monthly grids of terrestrial surface climate.

Often, A. and Stabbetorp, O.E. (2003): Landscape and biodiversity changes in a Norwegian agricultural landscape between 1960 and 2000 (in preparation).

Parmesan, C. and Yohe, G. (2003): 'A globally coherent fingerprint of climate change impacts across natural systems', Nature 421 pp. 37–42.

Parry, M.L. (ed) (2000): Assessment of potential effects and adaptation for climate change in Europe: The Europe Acacia Project, Jackson Environmental Institute, University of East Anglia, Norwich, UK. 320 pages.

Pauli, H., Gottfried, M., Dirnböck, T., Dullinger, S. and Grabherr, G. (2003): 'Assessing the long-term dynamics of endemic plants at summit habitats', In Nagy, L., Grabherr, G., Körner C. and Thompson, D.B.A. (eds), Alpine Biodiversity in Europe — a Europe-wide assessment of biological richness and change, Ecological Studies, Springer, Berlin, pp. 195–207.

Pauli, H., Gottfried, M. and Grabherr, G. (2001): 'High summits of the Alps in a changing climate. The oldest observation series on high mountain plant diversity in Europe', in Walther, G.-R., Burga, C.A. and Edwards, P. J. (eds), Fingerprints of climate change — adapted behaviour and shifting species ranges, Kluwer Academic Publishers, New York, pp. 139–149.

Preston, C.D., Telfer, M.G., Arnold, H.R., Carey, P.D., Cooper, J.M., Dines, T.D., Hill, M.O., Pearman, D.A., Roy, D.B. and Smart, S.M. (2002): The changing flora of the UK, London, DEFRA.

Root, T.L., Price, J.T., Hall, K.R., Schneider, S.H. Rosenzweig, C. and Pounds, J.A. (2003): 'Fingerprints of global warming on wild animals and plants', Nature 142, pp. 57–60.

Rustad, L.E., Campell, J.L., Marion, G.M., Norby, R.J., Mitchell, M.J., Hartley, A.E., Cornelissen, J.H.C. and Gurevitch, J. (2001): 'A meta-analysis of the response of soil respiration, net nitrogen mineralisation, and above ground plant growth to experimental ecosystem warming'. Oecologia 126. pp. 543–562. DOI 10.1007/s004420000544.

Sitch, S., Smith, B., Prentice, I. C., Arneth, A., Bondeau, A., Cramer, W., Kaplan, J., Levis, S., Lucht, W., Sykes, M., Thonicke, K. and Venevski, S., 2003. 'Evaluation of ecosystem dynamics, plant geography and terrestrial carbon cycling in the LPJ Dynamic Vegetation Model', Global Change Biology, 9, pp. 161–185.

Sykes, M. and Prentice, I.C. (1996): 'Climate change, tree species distribution and forest dynamics: A case study in the mixed conifer/northern hardwood zone of northern Europe', Climatic Change, 34, pp. 161–177.

Sykes M.T., Prentice, I.C. and Cramer, W. (1996): 'A bioclimatic model for the potential distributions of North European tree species under present and future climates'; Journal of Biogeography 23(2), pp. 203–233.

Tamis, W.L.M., Van 't Zelfde, M., and Van der Meijden, R., (2001): 'Changes in vascular plant biodiversity in the Netherlands in the twentieth century explained by climatic and other environmental characteristics', in Van Oene, H., Ellis, W.N., Heijmans, M.M.P.D., Mauquoy, D., Tamis, W.L.M., Berendse, F., Van Geel, B., Van der Meijden, R. and Ulenberg, S.A. (eds), Long-term effects of climate change on biodiversity and ecosystem processes, NOP, Bilthoven, pp. 23–51.

Theurillat, J.P., Felber, F., Geissler, P., Gobat, J.M., Fiertz, M., Fischlin, A., Kuipfer, P., Schlussel, A., Velutti, C. and Zhao, G.F. (1998): 'Sensitivity of plant and soil ecosystems of the Alps to climate change', in: Cebon, P., Dahinden, U., Davies, H.C., Imboden, D. and Jager, C.C. (eds.) Views from the Alps: Regional perspectives on climate change, MIT Press, Cambridge, UK.

Theurillat, J.P. and Guisan, A. (2001): 'Potential impact of climate change on vegetation in the European Alps: A review', Climatic Change 50, pp. 77–109.

Thomas, C.D., Cameron, A., Green, R.E., Bakkenes, M., Beaumont, L.J., Collingham, Y.C., Erasmus, B.F.N. et al., (2004): 'Extinction risk from climate change', Nature 427, pp. 145–148.

UN-ECE/FAO (2000): Forest resources of Europe, CIS, North America, Australia, Japan and New Zealand, Contribution to the global forest resources assessment 2000. UN, New York and Geneva, 445 pages.

Väre, H., Lampinen, R., Humphries, C. and Williams, P. (2003): 'Taxonomic diversity of vascular plants in the European alpine areas', in Nagy, L., Grabherr, G, Körner, C. (eds), Alpine biodiversity in Europe — A Europe-wide assessment of biological richness and change, Springer, Berlin, Germany: pp. 133–148.

Walther, G. R., Post, E., Convey, P., Menzel, A., Parmesan, C., Beebee, T. J. C., Fromentin, J. M. et al., (2002): Ecological responses to recent climate change. Nature 416: pp. 389–395.

Zhou, L., Tucker, C.J., Kaufmann, R.K., Slayback, D., Shabanov, N.V and Myneni, R.B. (2001): 'Variations in northern vegetation activity inferred from satellite data of vegetation index during 1981 to 1999', J. Geophys. Res., 106(D17): pp. 20069–20083.

Water

Kaspar, F. (2004): Entwicklung und Unsicherheitsanalyse eines globalen hydrologischen Modells. PhD thesis, University of Kassel, 139p, Kassel University Press (in press).

IPCC (2001a): Climate change 2001: The scientific basis, Cambridge University Press, Cambridge, UK.

Lehner, B., Henrichs, T., Döll, P. and Alcamo, J. (2001): EuroWasser: Model-based assessment of European water resources and hydrology in the face of global change. Centre for Environmental Systems Research, University of Kassel, Kassel World Water Series no. 5.

UNESCO, 1999. Discharge of selected river basins in the world. World Water Resources and their use: A joint product of the State Hydological Institute (SHI) and UNESCO, prepared by I.A. Shiklomanov. http://webworld.unesco. org/water/ihp/db/shiklomanov

Winsor, P., Rodhe, J. and Omstedt, A. (2001): 'Baltic Sea ocean climate: an analysis of 100 years of hydrographic data with a focus on the freshwater budget', Climate Research 18, pp. 5–15.

Agriculture

Beniston, M. (2004): 'The 2003 heatwave in Europe: A shape of things to come? An analysis based on Swiss climatological data and model simulations', Geophysical Research Letter, 31.

Carter, T.R. and Saarikko, R.A. (1996): 'Estimating regional crop potential in Finland under a changing climate', Agricultural and Forest Meteorology, 79, pp. 301–313.

European Commission (2002): European agriculture entering the 21st century. European Commission — Directorate-General for Agriculture.

FAO (2004): Food and Agriculture Organisation FAOSTAT data. http://www.fao.org

Hafner, S. (2003): 'Trends in maize, rice, and wheat yields for 188 nations over the past 40 years: a prevalence of linear growth', Agriculture, Ecosystems and Environment, 97, pp. 275–283.

Harrison, P.A., Butterfield, R.E. and Orr, J.L. (2003): 'Modelling climate change impacts on wheat potato and grapevine in Europe', in: Downing, T.E., Harrison, P.A., Butterfield, R.E. and Lonsdale, K.G. (eds) Climate change, climatic variability and agriculture in Europe, Environmental Change Institute, Oxford, UK.

Hulme, M., Barrow, E.M., Arnell, N.W., Harrison, P.A., Johns, T.C. and Downing, T.E. (1999): 'Relative impacts of human-induced climate change and natural climate variability', Nature, 397, pp. 688–691.

IPCC — Intergovernmental Panel on Climate Change (2001b): Climate change 2001: impacts, adaptation and vulnerability.

JRC — Joint Research Centre (2003): Synthesis of the campaign 2002/2003 and start of the new 2003/2004 campaign. MARS Bulletin, 11, pp. 1–21.

Kimball, B.A., Mauney, F.S. Nakayama, F.S. and Idso, S.B. (1993): 'Effects of elevated CO_2 and climate variables on plants', Journal of Soil Water Conservation, 48, pp. 9–14.

Oldeman, R.L., Hakkeling, T.A. and Sombroek, W.G. (1991): World map of the status of human-induced soil degradation, International Soil Reference and Information Centre, Wageningen.

Olesen, J.E. and Bindi, M. (2002): 'Consequences of climate change for European agricultural productivity, land use and policy', European Journal of Agronomy, 16, pp. 239–262.

Pinter, P.J.J., Kimball, B.A., Garcia, R.L., Wall, G.W., Hunsaker, D.J. and LaMorte, R.L. (1996): 'Free-air CO_2 enrichment: responses of cotton and wheat crops', in: Koch, G.W. and Mooney, H.A.

(eds), Carbon Dioxide and Terrestrial Ecosystems, pp. 215–249. Academic Press, San Diego, USA.

Economy

EEA (2004): Mapping the impacts of recent natural disasters and technological accidents in Europe, Environmental issue report No 35, European Environment Agency, Copenhagen, Denmark.

IPCC (2001a): Climate change 2001: The scientific basis, Cambridge University Press, Cambridge, UK.

Munich Re (2000): Topics-annual Review of Natural Disasters 1999, Munich Reinsurance Group, Munich, Germany.

Wirtz, A. (2004): Naturkatastrophen in Europa, Münchner Rück, unpublished.

Human health

Campbell-Lendrum D.H., Prüss-Üstün, A. and Corvalan, C. (2003): 'How much disease could climate change cause?', in: McMichael A.J. et al., (eds): Climate change and health: risks and responses, Geneva, Switzerland, World Health Organisation.

Curriero F. et al. (2002): 'Temperature and mortality in 11 cities of the eastern United States', American Journal of Epidemiology, 155, pp. 80–87.

Daniel M, Kríž, B. (2002): Tick-borne encephalitis in the Czech Republic: I. Predictive maps of Ixodes ricinus tick high-occurrence habitats and a tick-borne encephalitis risk assessment in Czech regions; II. Maps of tick-borne encephalitis incidence in the Czech Republic in 1971–2000. Project 1420cCASHh EVK2 — 2000–2002.

EEA (2004): Mapping the impacts of recent natural disasters and technological accidents in Europe, Environmental issue report No 35, European Environmental Agency, Copenhagen, Denmark.

Empereur-Bissonet (2004): Health impacts of the 2003 heat wave in France, Report on the WHO meeting on 'Extreme weather and climate events and public health' Bratislava, Slovakia.

Faunt, J.D., Wilkinson, T.J., Aplin, P., Henschke, P., Webb, M. and Penhall, R.K. (1995): 'The effete in the heat: heat-related hospital presentations during a ten day heat wave. Australia and New Zealand', Journal of Medicine, 25, pp. 117–120.

IPCC (2001a): Climate change 2001: The scientific basis, Cambridge University Press, Cambridge, UK.

IPCC (2001b): Climate change 2001: Impacts, adaptation and vulnerability, IPCC WGII report, Cambridge University Press, Cambridge, UK.

IVS (2003): Impact sanitaire de la vague de chaleur en France survenue en août 2003, Rapport d'étape, 29 août 2003. Saint-Maurice, Institut de Veille Sanitaire.

Jaenson T.G.T., Tälleklint, L. Lundqvist, L., Olsen, B., Chirico, J. and Mejlon, H. (1994): 'Geographical distribution, host associations, and vector roles of ticks (Acari: Ixodidae, Argasidae) in Sweden', J Med Entomol (2), pp. 240–56.

Jendritzky, G., Bucher, K. and Bendisch, F. (1997): Die Mortalitätsstudie des Deutschen Wetterdienstes. Annalen der Meteorologie, 33, pp. 46–51.

Jendritzky, G., et al. (2000): 'Atmospheric heat exchange of the human being, bioclimatic assessments, mortality and heat stress', International Journal of Circumpolar Health, 59, pp. 222–227.

Katsouyanni, K., Pantazopoulu, A., Touloumi, G., Tselepidaki, I., Moustris, K., Asimakopoulos, D., Poulopoulou, G. and Trichopoulos, D. (1993): 'Evidence of interaction between air pollution and high temperatures in the causation of excess mortality', Architecture and Environmental Health, 48, pp. 235–242.

Katsouyanni, K., Trichopoulos, D., Zavitsanos, X. and Touloumi, G. (1988): 'The 1987 Athens heatwave', Lancet, 573.

Kovats, S., Ebi, K.L. and Menne, B. (2003): Methods of assessing human health vulnerability and public health adaptation to climate change, WHO, WMO, Health Canada, UNEP.

Kunst, A.E., Looman, C.W.N. and Mackenbach, J.P. (1993): 'Outdoor air temperature and mortality in the Netherlands: a time–series analysis', American Journal of Epidemiology, 137, pp. 331–341.

Lindgren E, Tälleklint, L. and Polfeldt, T. (2000): 'Impact of climatic change on the northern latitude limit and population density of the disease-transmitting European tick, Ixodes ricinus', Environmental Health Perspectives 2000; 108(2), pp. 119–23.

Martens, W.J.M. (1997): 'Climate change, thermal stress and mortality changes', Social Science and Medicine, 46, pp. 331–344.

McMichael, A.J. and Kovats, R.S. (1998): Assessment of the impact on mortality in England and Wales of the heatwave and associated air pollution episode of 1976, Report to the Department of Health, London School of Hygiene and Tropical Medicine, London, UK.

Rizzoli A, Merler, S., Furlanello, C. and Genchi, C. (2002): 'Geographical information systems and bootstrap aggregation (bagging) of tree-based classifiers for Lyme disease risk prediction in Trentino, Italian Alps', J Med Entomol, (3), pp. 485–92.

Rooney, C., McMichael, A.J., Kovats, R.S. and Coleman, M. (1998): 'Excess mortality in England and Wales, and in Greater London, during the 1995 heatwave', Journal of Epidemiology and Community Health, 52, pp. 482–486.

Sartor, F. *et al.* (1995): 'Temperature, ambient ozone levels, and mortality during summer 1994 in Belgium', Environmental Research, 70, pp. 105–113.

Tälleklint, L. and Jaenson, T.G.T. (1998): 'Increasing geographical distribution and density of Ixodes ricinus (Acari: Ixodidae) in central and northern Sweden', J Med Entomol, 4, pp. 521–526.

WHO-ECEH (2003a): Extreme weather events and human health, Third intergovernmental Preparatory Meeting; Evora 2003.

WHO-ECEH (2003b): Climate change and human health risks and responses, Geneva, Switzerland.

WHO-ECEH (2004): Heatwaves: risks and responses. Health and global environmental changes, Series No. 2, Geneva, Switzerland.

4. Adaptation

Klein, R.J.T. and Tol, R.S.J. (1997): Adaptation to climate change: options and technologies, An Overview Paper. No. E-97/18. Institute for Environmental Studies Publication, Vrije Universiteit, Amsterdam, Netherlands.

Willows, R.I and Connell, R.K. (eds) (2003): Climate adaptation: Risk uncertainty and decision-making, UKCIP Technical Report. UKCIP, Oxford, UK.

5. Uncertainties, data availability and future needs

EEA (2002): Proposed set of climate change state and impact indicators in Europe. Technical Report, European Environmental Agency, Copenhagen, Denmark.

GCOS (2003): The second report on the adequacy of global observing systems for climate in support of the UNFCCC, report no. GCOS-82.

IPCC (2000): Special Report on Emissions Scenarios, Cambridge University Press, UK.

Kaspar, F. (2004): Entwicklung und Unsicherheitsanalyse eines globalen hydrologischen Modells. PhD thesis, University of Kassel, 139p, Kassel University Press (in press).

Theurillat, J.P., Felber, F., Geissler, P., Gobat, J.M., Fiertz, M., Fischlin, A., Kuipfer, P., Schlussel, A., Velutti, C. and Zhao, G.F. (1998): 'Sensitivity of plant and soil ecosystems of the Alps to climate change', in: Cebon, P., Dahinden, U., Davies, H.C., Imboden, D. and Jager, CC. (eds): Views from the Alps: Regional perspectives on climate change. MIT Press, Cambridge, UK.

European Environment Agency

Impacts of Europe's changing climate, An indicator-based assessment

Luxembourg: Office for Official Publications of the European Communities

2004 — 107 pp. — 21 x 29.7 cm

ISBN 92-9167-692-6

Price (excluding VAT) in Luxembourg: EUR 15.00